T0344493

Solving Problems of Simple Structural Mechanics

Solve problems in elementary structural mechanics thoughtfully and efficiently with this self-contained volume. Cover the basics of structural mechanics and focus on simple structures, truss frameworks, beams and frames, design choices and deformity. Carefully interrogate underlying assumptions for efficiencies in working out whilst expounding fundamental principles for a consistent understanding. Connect the practical world of indeterminate structures to their analysis, to underline benefits they impart to the latter: that certain analytical methods provide a wealth of efficient solutions for problems of indeterminate structures compared to determinate ones. Celebrate the beauty of analytical indeterminacy and its relationship to practical structures. Perfect for students invested in structural mechanics.

Keith Alexander Seffen leads internationally recognised research into shape-changing structures in the Advanced Structures Laboratory at Cambridge, which he co-founded.

Solving Problems of Simple Structural Mechanics

KEITH ALEXANDER SEFFEN

University of Cambridge

CAMBRIDGE
UNIVERSITY PRESS

CAMBRIDGE
UNIVERSITY PRESS

University Printing House, Cambridge CB2 8BS, United Kingdom

One Liberty Plaza, 20th Floor, New York, NY 10006, USA

477 Williamstown Road, Port Melbourne, VIC 3207, Australia

314–321, 3rd Floor, Plot 3, Splendor Forum, Jasola District Centre, New Delhi – 110025, India

103 Penang Road, #05–06/07, Visioncrest Commercial, Singapore 238467

Cambridge University Press is part of the University of Cambridge.

It furthers the University's mission by disseminating knowledge in the pursuit of education, learning, and research at the highest international levels of excellence.

www.cambridge.org
Information on this title: www.cambridge.org/9781108843812
DOI: 10.1017/9781108920131

First published 2022

A catalogue record for this publication is available from the British Library.

Library of Congress Cataloging-in-Publication Data
Names: Seffen, K. A. (Keith A.), author.
Title: Solving problems of simple structural mechanics / Keith Alexander Seffen,
    Department of Engineering, University of Cambridge, Trumpington Street, Cambridge.
Description: New York : Cambridge University Press, 2022. |
    Includes bibliographical references and index.
Identifiers: LCCN 2021017134 (print) | LCCN 2021017135 (ebook) |
    ISBN 9781108843812 (hardback) | ISBN 9781108920131 (epub)
Subjects: LCSH: Structural engineering–Mathematics.
Classification: LCC TA640 .S44 2021 (print) | LCC TA640 (ebook) | DDC 624.1–dc23
LC record available at https://lccn.loc.gov/2021017134
LC ebook record available at https://lccn.loc.gov/2021017135

ISBN 978-1-108-84381-2 Hardback

Contents

Preface

This book is for students already invested in Structural Mechanics. They know about forces and moments, and couples from pairs of applied forces. They understand the concepts of equilibrium, compatibility and stiffness; of how beams and pin-jointed frameworks differ in their constituent behaviour; of the nature of supports; and of the concept of statical equivalency.

I reflect the usual gradation in complexity and form, moving from bodies to bars, to cables, and to beams, and sometimes, mixing them up. I consider different metrics of structural design: of safe loading, of failure by plastic collapse and buckling, of cross-sectional limitations and joint design, for example.

In particular, I focus on how to think about (and solve) 'Structures' problems *better*. Formulaic analysis methods are not overly represented because they can be applied without understanding fully how they work. Instead, I present a dialogue of how solving proceeds, collated as short chapters of worked examples – without the usual repetitive exercises at their ends.

I apologise for my greyscale figures. I am a child of the hard-copy age where colour was taxed, limiting that part of my presentation still. I dispense with denoting vectors in boldface because the direction of quantity is always implied. Parameters that vary are italicised, but labels are roman and upright; 'A' can be a point in an area '*A*'. Greek letters typically denote fundamental quantities or dimensionless groups, but not always.

Purpose

I consider mostly *statically indeterminate* (or indeterminate, alone) structures but not exclusively, since despite more solving effort (because with equilibrium there are also geometrical compatibility and material laws, our three *imperatives*, to include), the scope for more efficient, more confident solutions is broadened. Also, because most practical structures are indeterminate.

For example, indeterminacy is often couched in terms of extra statical unknowns from 'too many' members or supports. Furthermore, a determinate structure (and thus a soluble one by equilibrium alone) can be wrought by subtracting certain of their number from the original indeterminate structure, which we then label to be *redundancies*.

But indeterminacy is a function of how the structure is built *entirely*: no particular member or support naturally identifies as being redundant. We can rightly set *any* statical quantity to be redundant in our analysis, especially if less working out follows – when finding elastic displacements via *Virtual Work*, for example.

Many good textbooks, however, insist that the imperatives, conceptually, stand alone from each other, which is reinforced by their sequential employment during solution. That equilibrium only depends on the initial geometry despite the distortions which follow, with the material obediently furnishing a governing linear connection between them. Of course, adhering to small displacements is one reason.

But it is precisely because they are linked during solution that indeterminacy allows us to probe deeper into fundamental behaviour. For example (again), mechanistic and thus significant (and potentially disastrous) departures from initial geometry can combine favourably with redundancy, to suggest a new type of *non-linear* stiffening overall as distortions accrue; see Chapter 4.

During plastic design, material ductility is mandatory, in order that any viable equilibrium solution, *i.e.* one of our choosing, expresses a safe loading. Put another way, we can choose the *wrong* equilibrium solution in view of compatibility yet achieve a safe working, which is anathema to our sense of engineering precision.

Compensating for our apparent error is wrought by the action of indeterminacy, specifically, by the plastic deformation being able to *redistribute* elsewhere in the structure because its material *is* ductile. In trade-off, the cross-section has to increase in size which, however, increases the safety factor for our structure.

Indeterminacy thus cultivates analytical advantage and kinematical insight. It should not be portrayed as the bane of structural simplicity, where solving determinate cases is largely an instructive exercise. I hope my examples provide suitable demonstration and a different insight into solving Structures. Their content is now described.

Layout

The following chapter quickly revises some key concepts: geometry and distortion, the *generalised* Hooke's laws for bars and beams, and well-known energy methods. There are then 21 chapters in 6 parts.

Simple Structures

The first structures are rigid bodies acted upon by frictional forces in Chapter 1. All examples have one statical unknown; and when Columb's Inequality ($F \leq \mu N$) is satisfied at the point of slipping, each system becomes determinate. Their equilibrium geometry can also be solved by graphical means – more quickly so, for the examples chosen.

Chapter 2 deals with the displacements of rigid bodies supported on elastic springs or by fluid buoyancy. These reaction forces depend on the very displacements they induce, and the deformed geometry must feature to define them; in effect, these forces assume a constitutive character. Nevertheless, by maintaining small displacements, overall equilibrium can be assessed via the initial layout.

Geometry and equilibrium come together again in Chapter 3 for loaded cables, which are infinitely flexible; their deformed equilibrium is governed similarly to beams.

Truss Frameworks

The next three chapters (Part II) deal with elastic truss frameworks.

We deploy *Maxwell's Rule* in Chapter 4 to calculate the number of redundancies. This rule is a statement of absolute rigidity, derived from counting the bars, joints and supports. A positive tally expresses the degree of indeterminacy, and *vice versa* for the number of mechanistic motions. For certain truss layouts, indeterminacy and mechanistic action can become conflated, where their stiffness 'emerges' only when there is deformation; Maxwell's Rule is modified accordingly.

Elastic bar extensions are calculated from the method of Virtual Work in Chapter 5. Because indeterminacy promotes infinitely many equilibrium sets of bar tensions, we can explore their variety in order to enhance the Virtual Work process. Several truss examples are presented, including one with *mis-fitting* bars.

Truss design by the Lower Bound method is considered in Chapter 6. The method is rarely applied to trusses, but in this example the exact *elasto-plastic* response is soluble, in order to highlight the method's efficacy, in particular the role played by the members' ductility in redistributing the loading capacity during permanent yielding of the material.

Beams and Frames: Character

Characterising beam and frame behaviour is then garnered in the three chapters of Part III.

Bending moment and shear force diagrams are their equilibrium signatures. In Chapter 7, we think about the loads themselves, especially about point loadings – in isolation; what happens *away* from loads is equally important for drawing diagrams properly. The sign convention, which commands little mention elsewhere, is strictly enforced at all times, and drawing diagrams becomes an holistic, assured exercise.

The level of redundancy in beams (and frames) is determined in Chapter 8 by *ad hoc* counting procedures based on inserting pins or cutting the structure. Furthermore, using symmetry (and anti-symmetry) concepts in Chapter 9, we can declare certain statical quantities to be zero; the arguments are also developed for kinematical quantities.

We limit ourselves to singly symmetrical/anti-symmetrical layouts, where the midline performance is key. Both halves of the structure are compared by *flipping* or *spinning*, where contradictory (and hence zero) parameters can be identified.

Beams and Frames: Analysis

The techniques above then contribute to our analysis of indeterminate structures (Part IV). Our analytical basis is the *Force Method*, which separates the structure into constituent determinate parts, sharing equal and opposite redundant forces and moments. Compatibility of their corresponding deflections furnishes a complete solution.

We therefore revise determinate beams and cantilevers in Chapter 10, making use of *polynomial functions* for their exact displacements. Particular layouts of initial geometry and loading lead to a list of *standard case* results, which underpin the method of *Deflection Coefficients* in Chapter 11 for solving indeterminate cases.

The stiffness of a simply-supported beam loaded by a couple is found in Chapter 12. Despite its determinate nature, the analysis is somewhat involved. The beam is then modified – and made redundant – by adding another roller support, which, importantly, is collocated with the couple. The new stiffness is much easier to calculate even though indeterminacy would prepare us for more effort. Furthermore, the bending moment performance around the new support enables us to think about the performance of a more detailed junction, of two or more beams meeting.

Design Choices

Making informed choices for the design of a structure then follow in Part V.

We recall first in Chapter 13 that pin-joints are an idealisation, yet many rigidly connected frameworks are treated as pin-jointed. We compare two identical layouts of a simple arch with rigid and pinned joints, loaded by point forces to promote axial forces as well as bending. As their member slenderness increases, we find a diminishing rigidity for all joints.

Chapter 14 introduces short-cuts for calculating the second moment of area for symmetrical cross-sections in which the neutral axis is the usual major (or minor) axis. When the applied bending moment is no longer parallel to either axis, the corresponding neutral axis is altogether different from the bending direction; calculating the elastic stiffness is no longer straightforward.

Optimal structural performance is then explored in two ways. First, at the level of cross-section, by comparing bending and torsional stiffness; the latter prepares us for analysis of 'pseudo' three-dimensional frames – *grillages*. Second, in terms of structural stiffness and strength from transverse loading alone, in order to highlight the 'natural' limits of cross-sectional proportions either way.

Establishing the strength of a cross-section takes place in Chapter 15. A Lower Bound approach permits any viable equilibrium solution, which may reduce our working with marginal sacrifice in the accurate strength. The interaction between multiple stress resultants also relates to joint detailing. Further Lower Bound demonstration for designing the diverse supports of multi-span beams is given in Chapter 16.

Analysis of indeterminate grillages is introduced in Chapter 17. Bending and torsion (and shearing) are automatically coupled because of out-of-plane loading, to a degree prescribed by the grillage layout and nature of its joints. We explore this coupling character by solving a variety of examples (using, in part, the method of Deflection Coefficients).

Deliberately Deformed

The role of deformed geometry is then celebrated in the final part – Part VI.

For collapse of an indeterminate beam or frame, we determine the least number of hinges required and their positions in Chapter 18, and thence the family of collapse modes. General collapse motions are calculated from the method of *Instantaneous Centres* for multiple, interacting loads; and the correct mechanism ultimately correlates with the best Lower Bound equilibrium solution.

Beam *buckling* in Chapter 19 focuses on the mathematical nature of the solution and the apparent paradox that displacements remain unresolved even though they must feature in the formulation; and that buckling for actual *imperfect* structures is a misnomer. Chapter 20 considers more elaborate buckling cases where the nature of the governing loads changes dramatically with deformation.

Finally, we consider the effect of heating a material upon the structural response in Chapter 21. The inclusion of temperature effects at a material level is straightforward, but the outcome structurally is more complex. The primary example is a bimetallic strip, whose performance *deliberately* celebrates displacements *increasing* when most structural design does not.

Acknowledgements

Writing a book about solving Structures problems inevitably echoes examples from other books; I aim, nevertheless, to express my own solutions – as this book purports. A few examples have been directly reworked from established texts and are referenced locally. Some examples have been taken directly, and inspired indirectly, from Part I of the Engineering Tripos at the University of Cambridge, with original figures redrawn and their solutions altogether reworked. Their inventors from the Structures Group are collectively credited but noting in particular Professors Chris Calladine (emeritus), Jacques Heyman (emeritus), Chris Burgoyne (emeritus), Simon Guest, Cam Middleton, Sergio Pellegrino (now at Caltech), Janet Lees, Allan McRobie, and Dr Chris Morley (retired). I have devised the rest either as part of my own Tripos lecturing duties or in writing this book. If I have been negligent in my credits (and gratitude) elsewhere, it has been unintentional.

if second-order terms are kept, the new length is $L(1 + (1/2) \cdot \theta^2)$, similar to the hypotenuse expression. There is thus an apparent axial strain equal to $\theta^2/2$.

This is not physically possible but arises analytically because $L\theta$ has been assumed to act normal to the original bar axis. Movement of the rotating end on a strict circular path also produces a second-order axial component of displacement, equivalent to an equal and opposite (and thence negating) strain compared to above.

These second-order effects in the kinematics of deformation are an example of *geometrical non-linearity*. They are, for most problems, negligible, but in certain cases they provide valuable insight when first-order effects are remiss. Invoking them, however, requires careful thought about how they are formulated, as we have just seen.

Curving

The curvature of a circle of radius R is obviously $1/R$; or $2\pi/2\pi R$, the total angle enclosed (or subtended) by the circle, divided by the circumference. More generally, a portion of circumference with arc-length s, equal to $R\theta$, sets $1/R = \theta/s$ and the curvature as the ratio of the local subtended angle to arc-length.

This definition can be contracted to elemental values when the curvature (and R) varies along the 'curve'. Denoting by κ:

$$\kappa = \lim_{\delta s \to 0} \frac{\delta\theta}{\delta s} = \frac{d\theta}{ds}. \tag{0.3}$$

When beams deflect, they engage displacements *transverse* to their original stress-free layout. Geometrical continuity demands curving of the beam, and thus displacements are related to curvature and curvature itself to the structural response of the beam. We should also imagine the beam to be slender, *i.e.* very thin compared to its length, as if condensed into a line for the beam axis.

An orthogonal coordinate system can be superposed conveniently onto an originally straight beam with x along and v normal for displacements. The gradient of the deflected shape is now dv/ds along an *intrinsic* coordinate s on the displaced beam.

Over a small deflected element of original length δx, the displacements change by δv, and Pythagoras sets $\delta s^2 = \delta v^2 + \delta x^2$. Dividing through by δx, using the Binomial Theorem, and observing the limit, we find

$$\frac{ds}{dx} \approx 1 + \frac{1}{2}\left(\frac{dv}{dx}\right)^2. \tag{0.4}$$

When dv/dx is small, the second term above is negligible compared to unity, setting $\delta s \approx \delta x$ and $dv/ds \approx dv/dx$. This is known as the shallow gradient assumption, which predicates small displacements for moderate length, slender beams.

We can now replace θ by dv/dx and δs by δx in the original definition of curvature, Eq (0.3), to give

$$\kappa = \frac{d}{dx}\left(\frac{dv}{dx}\right) = \frac{d^2v}{dx^2}. \tag{0.5}$$

Author's Note

Straining

Often we have to deal with relatively small changes in geometrical quantities because of simplifying assumptions. Consider, for example, a narrow right-angled triangle. We find the length of the hypotenuse to be $\sqrt{L^2 + a^2}$ where a is the smallest side-length; or $L\sqrt{1 + (a/L)^2}$.

We make further progress using the Binomial Theorem, which states,

$$(1 + x)^n = 1 + nx + \frac{n(n-1)}{2!} \cdot x^2 \ldots \ldots \tag{0.1}$$

Thus, our hypotenuse length becomes

$$L\left[1 + \frac{1}{2}\left(\frac{a}{L}\right)^2 - \frac{1}{8}\left(\frac{a}{L}\right)^4 \ldots\right] \approx L\left[1 + \frac{1}{2}\left(\frac{a}{L}\right)^2\right] \tag{0.2}$$

when a is much smaller than L, and a/L much less unity: for $x \ll 1$, we may set $(1 + x)^n$ equal to $1 + nx$.

When a rigid bar of length L pivots in plane by a small angle θ about one end, the other experiences a small normal displacement $L\theta$. If the bar is inclined at a general angle α to the horizontal, the vertical and horizontal displacement components are $L\theta \cdot \cos\alpha$ and $L\theta \cdot \sin\alpha$, respectively: or, incidentally, $L\cos\alpha \cdot \theta$ and $L\sin\alpha \cdot \theta$, the same. The components multiply the absolute rotation by the *projections* of the bar length onto the required directions.

Now consider an elastic bar, which also extends by a small length e. The in-plane displacement components of one end relative to the other are now e and $L\theta$. Clearly, the axial strain of the bar, e/L, is unaffected by the rotation, no matter its size.

If the strain is computed instead from the change in bar length according to the displacement components of its end, we first observe a new length of $\sqrt{(L+e)^2 + (L\theta)^2}$. Using the Binomial Theorem and retaining terms up to second order, we have $L(1 + e/L + (1/2) \cdot [(e/L)^2 + \theta^2])$.

Discounting the much smaller squared terms sets the length to be $L(1 + e/L)$, hence, our expected strain. Even though the orthogonal displacement components may be of similar size, the axial strain is garnered only from e, the component along the bar.

However, when e is negligible compared to $L\theta$, we can make two claims about the strain. Either there is no strain because there is only pure rigid body rotation or,

The *direction* of curvature at a given position on the beam axis can be specified by where the centre of curvature lies relative to the axis, *i.e.* the origin of the local *radius* of curvature, much like the centre of our first circle.

For example, let the displacements be described by $v = Ax^2/2$, an upwardly curving parabola. Equation (0.5) returns a constant curvature, $\kappa = A$, and thus a centre of radius located $1/A$ *above* the original beam line. For this, our definition of positive curvature, Eq. (0.5) remains; if not, the right-hand side is multiplied by -1.

Stiffness and Hooke

At a material level, elastic stress is linearly related to strain by the Young's Modulus, E, through Hooke's Law. This is our *constitutive law* for *direct* behaviour, when the stress acts normal to an exposed cross-section.

Shearing, on the other hand, acts tangentially with an equivalent modulus, G. For isotropic materials, $G = E/2(1+v) \approx 0.38E$ when the Poisson's Ratio, v, equals 0.3 (as in most Engineering metals).

Multiplying a uniformly distributed stress, σ, by a cross-sectional area, A, we have an axial force, F, say. Applying F as equal and opposite forces to the ends of an elastic bar, it extends relatively and axially by e. The axial strain, ϵ, is e/L given an unstressed bar length, L. From Hooke's Law, $\sigma = E\epsilon$:

$$\frac{F}{A} = E \cdot \frac{e}{L} \quad \rightarrow \quad F = \frac{EA}{L} \cdot e. \tag{0.6}$$

The force is in linear relation to the extension, with a constant of proportionality now equal to our bar's axial stiffness. Reflecting the character of constitutive behaviour at a structural level is said to express a *generalised* Hooke's Law.

Note that we have invoked three *imperatives*. We have an equilibrium statement $F = \sigma A$ and a compatibility statement $e = \epsilon L$, bound together by the original material Hooke's Law. The internal force represents an aggregated stress performance, or a *stress resultant*. The uni-dimensional nature of Eq. (0.6) also suggests no difference between a physical bar, with area, and a theoretical elastic line of the same stiffness.

We proceed in the same way for describing beam bending. Before, we said that the beam *axis* becomes curved, but what is this axis? We can select any given horizontal plane within the original (and horizontal) beam for now (which in side view manifests as a line).

Successive planes above and below this must curve differently, in order to preserve their separation when there are no stresses and thus no straining through the depth. Imagine now a uniformly curved portion of beam subtending θ on the reference plane. Another plane, originally y above, has a current length $(y + R)\theta$, where $\kappa = 1/R$ is the reference plane curvature. The plane above becomes strained by an amount $y\kappa$, comparing its lengths before and after curving.

The linear variation of strain with height y leads to the familiar assumption that *plane sections remain plane*, and a given cross-sectional plane rotates during curving (about a line on the horizontal reference plane).

From Hooke's Law, the axial stresses are linear too and produce a turning effect about the reference plane. Their aggregation is another stress resultant, our *bending moment*. Its formal calculation considers an elemental force, $\sigma \cdot w(y)\delta y$, where $w(y)$ is the current width of cross-section at height y; hence $M = \int \sigma w(y) y \, dy$

Replacing σ by $E\epsilon$ with $\epsilon = y\kappa$, we find $M = E\kappa \int y^2 w(y) \, dy$. The integral term is the *second moment of area* (from the second-degree variation), abbreviated to I, setting $M = EI\kappa$, our generalised Hooke's Law for beam bending: of a stress resultant, its commensurate deformation, and a stiffness term made of a material constant and a geometrical property of the cross-section.

We also have a uni-dimensional expression, analogous to curving of an elastic line with EI as its *bending stiffness*; I, of course, is calculated from the actual cross-section.

To perform its integration, we must know the integration limits for y, which depend on the position of the reference plane. There is no axial force applied, demanding axial equilibrium of the stresses from bending.

We already know each elemental force, and formally integrating them over the cross-section sets $\int \sigma w(y) \, dy = 0$: or, $E\kappa \int y w(y) \, dy = 0$. Writing $w(y)\, dy$ as an elemental area, dA, we therefore observe $\int y \, dA = 0$, which is how we locate the *centroid* of the cross-section in the y direction: the reference plane passes through it.

The axial stresses can now be determined from cross-sectional properties. Given that $\epsilon = \sigma/E$ also equals $y\kappa$, with $\kappa = M/EI$, we can re-arrange and obtain $\sigma = My/I$.

Energy Methods

Energy methods ultimately describe the equilibrium or kinematic behaviour of a structure, usually from applying a work 'recipe' for the relevant parameters. For example, the method of *Virtual Work* is a statement of internal energy stored *vs* external effort, where, for a truss:

$$\Sigma_{\text{joints}} W \cdot \Delta = \Sigma_{\text{bars}} T \cdot e. \tag{0.7}$$

The external loads applied to pin-joints are W, and joint displacements are Δ; bar tensions are T and axial extensions, e. Note that W and T are always in equilibrium, and Δ forms a *compatible set* with e in which the joints displace in exact accordance with how the bars extend and rotate.

The energetic terms come from linear elastic behaviour, where normally we expect a 'half' pre-multiplying both sides (even though it would cancel across). This is because Eq. (0.7) is about the effect of *perturbations* – small changes to the configuration of the truss system. For example, if a loaded joint is displaced a little more, the value of applied load does not change.

The Virtual Work performed is therefore small amounts of surplus energy and effort from imposing extra, or *virtual*, loads and displacements *etc*. The operation of Eq. (0.7) also decouples equilibrium from compatibility: our virtual equilibrium set applies to the actual kinematics, as do any virtual kinematics upon the real equilibrium set.

Virtual Work can also be used to provide new theorems[1] such as the *Lower* and *Upper Bound Theorems*, whose purpose guarantees different outcomes for a loaded structure in view of failure.

If we are able to find *any* equilibrium solution that nowhere violates material yielding, the structure will stand safely; but it will collapse if we can postulate a mechanism compatible with how members fail. These seem obvious statements but are non-trivial to prove.

When the structure is originally indeterminate, there is more than one solution for both, with different loading values. They overlap exactly when the values are the same, and a safe structure is about to collapse.

We are not, however, obliged to find the correct solution, which may reduce our work considerably but which may lead to conservative behaviour. Consequently, a Lower Bound solution always gives a safe loading value that can potentially be increased, and an Upper Bound collapse load can potentially reduce.

---

[1] J Heyman, *Basic Structural Theory*, Appendix B, Cambridge University Press, 2008.

Part I

Simple Structures

1 Mastering Friction

Friction is both welcome and unwelcome in Engineering. Negligible friction between moving parts leads to low losses through heat and sound, giving high transducer efficiencies; high friction can give excellent grip and contact between surfaces when needed, as in clutch plates, road tyres *etc*. Understanding friction is therefore central to a good Engineering performance.

Its treatment usually begins with how several, often prismatic bodies interact and maintain statical equilibrium. When friction is insufficient, there is usually slippage between them or, in the extreme, a loss of contact altogether. The initiation of this otherwise dynamic phase can be viewed as a limiting quasi-static problem without inertial forces.

Friction imparts to the problem a *constitutive* statement in the sense of a relationship between forces and kinematics – in this case, of relative motion between bodies. Thus, we may write in addition to force and moment balances, a limiting *inequality* of the ratio of friction force to normal contact reaction, in order to test for slippage or not.

But consider a different viewpoint, of slippage from the outset. The inequality is always satisfied and the friction forces are uniquely related; or, the resultant of friction and normal forces is of both fixed size and fixed direction. There is now a single force pointing away from the direction of slippage, which, for the purposes of simple statics problems, can admit immediate information about the character of equilibrium without its explicit solution.

For example, consider two cases of equilibrium of a familiar heavy ladder of uniform mass standing on a horizontal floor and leaning against a vertical wall: when the wall is smooth and the floor is rough, and *vice versa* (Fig. 1.1). Let the coefficient of friction be μ, any friction force denoted by F, and normal reactions by N.

From the (planar) free-body diagram of the ladder by itself in Fig. 1.1(a), we may traditionally write two equations of force equilibrium and one of moment about a normal axis through its lower end, along with limiting friction:

$$N_1 = W \quad \text{(a)}, \quad F = N_2 \quad \text{(b)}, \quad N_2 L \sin\theta - W(L/2)\cos\theta = 0 \quad \text{(c)}, \quad F = \mu N_1 \quad \text{(d)}.$$
$$(1.1)$$

There are four unknowns (N_1, N_2, F, μ) in four equations, thus enabling a complete solution in terms of the layout specified by L and θ, and the self-weight, W. Equations (1.1(a) and (d)) tell us that $F = \mu W$, which substitutes for N_2 in Eq. (1.1)(b) and ultimately in Eq. (1.1)(c) where, after tidying up, we have $2\mu = \cot\theta$.

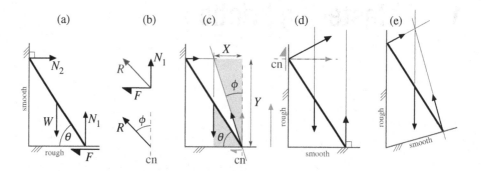

Figure 1.1 (a) External forces acting on a uniform inclined ladder. (b) Combining the normal reaction and friction force into a single frictional resultant, R, inclined at ϕ, the angle of friction at slippage, to the common normal (cn). (c) Reduction of (a) to three concurrent forces, with pertinent geometry. (d) Non-concurrent external forces acting on the same ladder when the floor is smooth and the wall is rough. (e) However, concurrent forces are possible when the floor from (d) is inclined.

This is a statement between the friction coefficient and layout, where the length and self-weight are not present. Of course, they set the absolute force levels, but the initial geometry ultimately dictates the equilibrium requirement. It also fits with our expectations of the ladder's stability: a larger angle causes $\cot\theta$ and hence μ to decrease, with a steeper ladder less prone to slippage.

Writing down all equations gives assurance but is inefficient – led, deliberately, by our imprecise question. If, as we proposed earlier, we combine F and N_1 into a single resultant, which we denote by R, it must be inclined to the common normal, the line of action of N_1, by angle ϕ (say), our *angle of friction* where $\tan\phi = \mu$ from limiting friction, and conveyed in a simple vector diagram, Fig. 1.1(b).

We have now reduced our problem to one with three force resultants by declaring *a priori* a known vector statement; and by reducing the number of unknowns, an easier solution beckons.

In working directly now with R, Eqs. (1.1a–d) no longer apply. We can write another set of equilibrium equations in terms of R, but this is tantamount to substituting the fourth equation into the remaining three. Instead, we pay attention to how these three forces are drawn in Fig. 1.1(c).

The normal contact force at the wall is horizontal, W is vertical, and the intersection of their action lines defines where R should pass through, for moment equilibrium about the same point. The equilibrium relationship now emerges from this geometry, where the height, Y, and half-width, X, between end points yield

$$\tan\phi = \frac{X}{Y}; \quad \tan\theta = \frac{Y}{2X} \quad \rightarrow \quad \tan\phi = \mu = \frac{\cot\theta}{2}.$$

This is a bolder, *quicker* solution, which empowers that for the second case. Figure 1.1(d) shows the same three forces where the smooth floor produces a vertical normal reaction parallel to W. No matter the inclination of R at the rough wall, we

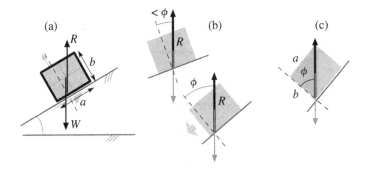

Figure 1.2 (a) External forces acting on a uniform block resting on a rough slope: the dashed line is the common normal. (b) Limiting friction and slippage when R becomes inclined at ϕ, now the slope angle of repose. (c) Limiting equilibrium with R at corner.

can never achieve force concurrency and possible statical equilibrium except for when all forces are parallel: ϕ will be equal to $90°$, but this is impossible for it sets μ to be infinity.

Our informal assessment of similar problems is also expedited by a graphical approach, for example, we see in Fig. 1.1(e) that if the smooth floor were itself inclined, equilibrium *becomes* possible because all of the action lines of forces can intersect.

We have proposed R to be inclined at ϕ for limiting friction, but if equilibrium is upheld for a smaller inclination, there is clearly no slippage. For example, a uniform block sits on a rough slope, which is progressively steepened until the block slips, as shown in Fig. 1.2(a).

The two resultant forces, W and R, must be collinear, with R becoming more inclined to the rotating common normal as the slope increases. Slippage occurs when R 'reaches' the required inclination of ϕ, which is also the angle of the slope, more commonly known as the *angle of repose*. This gives us a straightforward experimental check for the coefficient of friction between two surfaces.

1.1   Toppling *vs* Sliding

There is another limiting outcome for the block's repose in Fig. 1.2. As the slope steepens, R migrates towards the lowest point on the block, as shown in Fig. 1.2(c), and arrives there without slippage occurring provided $\tan \phi$ is greater than a/b. It cannot move outside the block, and further steepening leads to W and R separating. Moment equilibrium is now violated and the block will topple first before slipping.

The transition between slippage and toppling is therefore marked by R, already inclined at ϕ, passing through the block corner, which makes for a very precise relationship between the block size and μ in Fig. 1.2. The problem in Fig. 1.3 makes for their more convenient interaction, where the direction of toppling can also change.

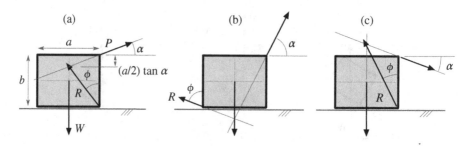

Figure 1.3 (a) Concurrent external forces acting on a uniform block about to topple forward due to force P inclined upwards. (b) Layout of forces just before the block topples backwards. (c) Their layout for toppling forward when P points downwards.

We have a horizontal block being towed to the right by a tensile force, P, applied to the top right corner, as shown in Fig. 1.3(a), and directed at positive angle α above the horizontal. There are three limiting toppling scenarios. First, P is directed upwards ($\alpha > 0$), and its line of action intersects that of W above ground. Since R resists the intended direction of movement, it always points backwards to the left at angle ϕ. Limiting moment equilibrium occurs when R is located at the front bottom corner, with the block tending to rotate forward.

For increasing α, P and W ultimately intersect below ground, but R cannot be inclined beyond $\phi = 90°$. It must be located instead at the rear bottom corner, as shown in Fig. 1.3(b), with the block rotating backwards and lifting off. Third, P points downwards ($\alpha < 0$) with R now again at the front corner, as shown in Fig. 1.3(c), as the block topples forward.

If we define ρ to be the aspect ratio of the block, a/b, the geometry of the concurrent forces in Fig. 1.3(a) reveals the following:

$$\tan \phi = \frac{a/2}{b - (a/2)\tan \alpha} \quad \rightarrow \quad \tan \phi = \frac{\rho}{2 - \rho \tan \alpha}. \tag{1.2}$$

Figure 1.3(c) also expresses this relationship when α takes negative values, so it is valid from $\alpha = -90°$ up to $\alpha = \arctan(2/\rho)$ when the denominator equals zero. At this value of $\alpha = \alpha^*$, we have $\phi = 90°$ with R horizontal. This marks the transition to the second case, where Fig. 1.3(b) can be used to show that $\tan \phi = \rho/(\rho \tan \alpha - 2)$ for $\alpha > \alpha^*$.

A compact *interaction diagram* of ϕ vs α expresses these theoretical limits in Fig. 1.4(a) by plotting the above equations around $\alpha = \alpha^*$. Sensibly, we see at either $\alpha = \pm 90°$ that $\phi = 0$, suggesting that, in theory, toppling always occurs first when the surface is smooth; after some rotation, of course, the block may slip as the configuration of forces changes, but that is a different problem.

In practice μ is fixed by the 'available' friction via $\tan \phi = \mu$. To now interpret the diagram, we first plot a horizontal line $\phi = \phi^*$, chosen to be 30°, with $\rho = 1$ for a square block. Where this line lies above the boundary curve, we have toppling; where

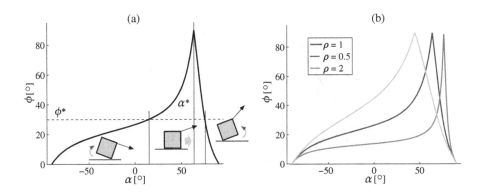

Figure 1.4 Limiting variation between toppling (above curve) and slippage (underneath) of a uniform block being towed along a rough floor. The angle of friction is ϕ, the aspect ratio of the block is ρ, and the towing force is inclined at α to the horizontal. A fixed coefficient of friction sets $\phi = \phi^*$ (dashed line); at α^* the direction of toppling reverses.

it lies below, the value of ϕ cannot meet the value set by the toppling requirement, so the block slips.

Reading from left to right, the block will topple forward, slip, then tip backwards, where the change in behaviour at different values of α will depend on ρ, which gives slightly different curves, as shown in Fig. 1.4(b).

1.2 Different Shapes

The ratio, ρ, tells us about the size of the contacting face relative to the height of the applied force. If we had a different shape of block, such as a triangle or parallelogram, then the form of previous limiting toppling equations remains the same.

On the other hand, a circular cylinder makes contact along a horizontal line, or a point if planar, giving us rolling instead of toppling as a limiting equilibrium scenario. In addition to being inclined at ϕ to the common (radial) normal, R is uniquely located at the contact point.

This further sets the geometry of solution and enables graphical solutions for cylinder problems with four forces, as we shall see. First, the cylinder of radius r and weight W in Fig. 1.5(a) is pulled over a rough step of height Y ($\leq r$) by a horizontal force, P, without slipping. We wish to find the minimum coefficient of friction and corresponding P.

As the cylinder commences overturning, floor contact is lost and the lines of action of the remaining three forces, R, W and P, intersect at the top of the cylinder, as shown in Fig. 1.5(a). Rather than specify the layout in terms of Y, we draw angle α inclined to the horizontal underneath the common normal, as shown in Fig. 1.5(b), which defines $\sin \alpha$ to be $(r - Y)/r$. We also note the horizontal distance X from $\cos \alpha = X/r$ which defines $X^2 = 2rY - Y^2$ using $\sin^2 \alpha + \cos^2 \alpha = 1$.

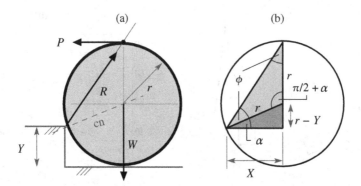

Figure 1.5 (a) Concurrent external forces acting on a uniform cylinder about to be pulled over a vertical step by a horizontal force P. (b) Angle of limiting friction, ϕ, and pertinent geometry.

Two triangles are now highlighted containing ϕ and α. The upper one is isosceles because two of its sides are radii: the obtuse angle must equal $\pi/2+\alpha$ from continuity of the vertical line through the lower triangle, which returns $2\phi+\pi/2+\alpha = \pi$ for the upper one, *i.e.* $\phi = \pi/4 - \alpha/2$. Knowing $\mu = \tan\phi$, the limiting state is expressed as follows:

$$\tan\phi = \tan(\pi/4 - \alpha/2) = \frac{1 - \tan(\alpha/2)}{1 + \tan(\alpha/2)} = \frac{\cos\alpha}{1 + \sin\alpha} \tag{1.3}$$

after using half-angle formulae for $\tan(\alpha/2)$. Furthermore, the right-hand side can be written in terms of the original geometry as $X/(2r - Y)$, and the force P is simply found by taking moments about the contact point, $P(2r - Y) = WX$, whence P.

The cylinder is now tethered horizontally to a rough slope of inclination α by a rigid cable, see Fig. 1.6(a). Equilibrium is maintained by three forces enclosing the highlighted triangle, where limiting friction sets $\phi = \alpha/2$. The cylinder tends to slip down the slope since R acts against it.

To counter this tendency, we apply a vertical upward force, V, at the most easterly point on the cylinder, as shown in Fig. 1.6(b), until R becomes inclined backwards to the common normal at $\phi = \alpha/2$. Such inclination defines where R and the cable tension, P, intersect, about which we take moments to yield limiting V in terms of W directly. If X is the distance from the intersection point to V, as shown in Fig. 1.6(b), then $V = W(X - r)/X$.

The cylinder also tends to slip up-slope when V is applied vertically *downwards* on the other side, as shown in Fig. 1.6(c), and the same four forces suggest a similar solution approach. From moment equilibrium, we immediately see that V is negative if R and P intersect to the left of V. Their intersection point must therefore lie to the right of V, with two possible outcomes. Either α is small enough so that ϕ can be equal to $\alpha/2$ to give limiting R and slippage, or R is less inclined than $\alpha/2$ without slippage.

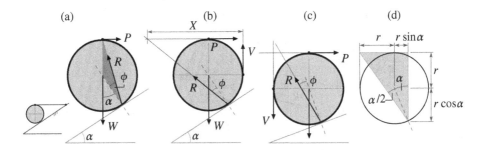

Figure 1.6 Uniform cylinder tethered horizontally to a rough slope. (a) Limiting friction with the cylinder slipping down the slope. (b) Slippage up the slope due to an extra, vertical force, V, applied upwards on the right side. (c) V applied downwards on the left side. (d) Geometry pertaining to (c).

The crossover occurs when the lines of action of R, V and P coincide, with Fig. 1.6(d) showing that $\tan(3\alpha/2) = (1 + \sin\alpha)/(1 + \cos\alpha)$ *i.e.* $\alpha = 0.42$ rad $= 24.3°$.

A larger α allows R, V and P to become very large without slippage, but obviously not W. To find the relationships between them, we could ignore W on grounds that it is much smaller and treat this case as a simpler three-force problem. We should not be surprised (or alarmed) that the friction forces can become infinitely large, notwithstanding local damage at the contact point: V *reinforces* the normal reaction, being generally opposed to it, which increases the friction force tolerance given $F \le \mu N$; in Fig. 1.6(b) the action of V diminishes N to bolster earlier slippage.

Since no slippage is possible no matter how large the forces involved, the configuration can become 'locked', and usefully so. For example, the wall-mounted 'paper-clip' shown schematically in Fig. 1.7(a) has a cylinder nestled inside a channel with one face inclined inwards to prevent the cylinder falling out under gravity.[1] The cylinder is rough with friction present at its contact lines.

The device works by feeding a paper sheet up through the gap between the vertical channel side and the cylinder, which accommodates by moving into the expanding space; if we now pull the sheet down or let it go, the cylinder rolls into place with contact between all surfaces, as shown, and the sheet is held. The channel is rigid, causing the sheet to tear under a large enough pulling force. In reality though, the channel is flexible and can bend outwards instead, thereby releasing the sheet.

If the weight of the cylinder can be neglected, only two equal and opposite forces, R, are applied to it during locking, as shown in Fig. 1.7(b). Their points of application are normal to the channel sides where contact is made, with one side inclined at angle θ to the vertical. The forces are also collinear, which dictates their orientations and thus inclinations to each common normal. These angles are both $\theta/2$, setting $\mu = \tan(\theta/2)$ as the minimum required coefficient of friction.

---

[1] F G J Norton, *Advanced Level Applied Mathematics*, Heinemann Educational Books, London, 1974, p. 282, qu. 16.

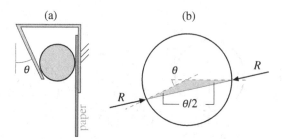

(a)   (b)

Figure 1.7 (a) Mounted paper clip/cylinder for holding sheets vertically. (b) Equal and opposite resultant forces on the cylinder.

1.3   Distributed Friction

When dealing with prismatic blocks, we have assumed friction forces to be concentrated. In practice, however, there is a normal contact pressure and thus a distributed frictional *intensity*. But because every contacting point experiences the same kinematical tendency, we may consider limiting behaviour in terms of their resultants – applied to/acting through the correct points, as the previous sections attest.

In the following example, slippage occurs in opposite directions *simultaneously* for a single body, which commands a different solution approach. The plan-view in Fig. 1.8(a) shows a narrow rod of uniform weight W and length L sitting on a rough horizontal plane; gravity acts normal to this plane. A force, P, is applied normal to the rod in the same plane at a point αL from one end, causing the rod to slip on the plane.

The rod is shallow in height and does not topple, and its width in plan is negligibly small compared to its length. Importantly, the rod does not translate uniformly (except when $\alpha = 1/2$: see later) but must also rotate initially for force and moment equilibrium. The direction of rotation is assumed to be anti-clockwise as shown, which stipulates a starting value of $\alpha = 1/2$ if P is to be positive for the same sense of rotation (up to a maximum value of $\alpha = 1$).

The point of rotation is generally located a distance βL from the same end as α, as shown in Fig. 1.8(b). The portion of rod before this point therefore slips backwards against P, and the rest forwards.

The out-of-plane contact pressure from gravity on the rectangular base of the rod is uniformly distributed. We can therefore divide the distribution of weight across the rotation point by length alone to give two normal out-of-plane reaction forces, respectively $(W/L) \cdot \beta L$ and $(W/L) \cdot (1 - \beta)L$. These act at the centre of each portion with corresponding friction forces $F_1 = \beta f$ and $F_2 = (1 - \beta)f$ in Fig. 1.8(c) after defining $f = W\mu$.

There are no left–right forces, so we resolve normally to the rod to find

$$P + F_1 - F_2 = 0 \quad \rightarrow \quad P = f(1 - 2\beta). \tag{1.4}$$

Taking moments about, say, the left end and noting the lever arm distances to the line of action of each force, we also find

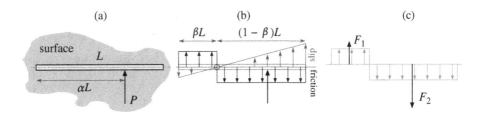

Figure 1.8 (a) Uniform rod on a horizontal plane (grey) acted upon by a normal force, P. (b) Rod slips on the plane by pivoting about a point βL from the left end, generating distributed, opposing frictional loading intensities. (c) Resultant friction forces acting though the centroid of intensities in (b).

$$P\alpha L + F_1 \frac{\beta L}{2} - F_2 \left[\beta L + \frac{(1-\beta)L}{2} \right] \quad \rightarrow \quad P\alpha = \frac{f}{2}(1-2\beta^2). \qquad (1.5)$$

Eliminating P between both equations returns a quadratic in β, which may be solved to give

$$(1-2\beta) = \frac{1}{2\alpha}(1-2\beta^2) \quad \rightarrow \quad \beta = \alpha \pm \sqrt{\alpha^2 - \alpha + 1/2}. \qquad (1.6)$$

The larger of these two roots sets $\beta > 1$; or, the rotation point ahead of P, which is impossible in the sense of anti-clockwise rotation: the smaller root only is valid. For example, when $\alpha = 1$, $\beta = 1 - \sqrt{0.5} \approx 0.3$, about 20% of the rod length from the centre (where $P/f \approx 0.41$).

Finally, when P is halfway and $\alpha = 1/2$, then β is found to be zero, giving a point of rotation at the left end of the rod. However, F_1 is also zero, with only F_2 present; the now symmetrical loading precludes any possible rotation of the rod. We can, of course, let P approach halfway in the limiting case, with solutions for β also approaching the left end: these are valid solutions as far as, but not including, the case of $\beta = 0$.

1.4   Static vs Kinetic Friction

When a body slips, its 'dynamic' coefficient of friction is marginally smaller than the static value before motion takes place: from our experiences of pushing a block on a surface, it is slightly easier to maintain slippage than to initiate it. This difference can also disrupt the symmetry of motion when two or more sliding bodies interact; and the example in Fig. 1.9 elegantly demonstrates this point.

A heavy horizontal rod rests initially on two cylinders asymmetrically positioned about the rod centre. Equal and opposite forces are then applied to each cylinder in order to initiate their approaching movement, as shown in Fig. 1.9(a.i). The cylinder on the right, being farthest away, exerts a smaller normal reaction on the rod compared to that on the left; both apply the same axial force inwards. The resultant frictional force,

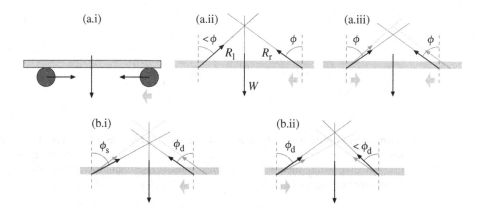

Figure 1.9 (a.i) Uniform rod resting horizontally on a pair of rough asymmetrical cylinders, which are then drawn towards each other; (a.ii) external concurrent forces acting on the rod initially; (a.iii) symmetrical configuration after slippage of the right cylinder. (b.i) Continued slippage of the right cylinder and forces: ϕ_d is the dynamic angle of friction compared to the (larger) static value, ϕ_s; (b.ii) reversal in slippage to the left cylinder.

R_r, is thus more inclined than R_l, and both are concurrent with the third force, W, the weight of the rod, for moment equilibrium.

Assume first that the coefficients of friction are the same and equal to μ (= $\tan \phi$). As the axial forces increase, R_r and R_l lean further away from their common normals, but R_r reaches ϕ first. The right cylinder therefore moves first to the left (quasi-statically), Fig. 1.9(a.ii). Since the inclination of R_r remains fixed, the intersection point of the three forces lowers during its movement, also making R_l more inclined. Eventually, the inclination of R_l reaches ϕ, giving a symmetrical layout of forces, Fig. 1.9(a.iii). The left cylinder can now slip, and both move together symmetrically at the same rate.

Different coefficients of friction do not affect the initial motion provided the right-side cylinder is appreciably off-centre. During slippage, $\mu = \mu_d$ with a corresponding ϕ_d from $\tan \phi_d = \mu_d$, with both parameters being smaller than their respective static values, ϕ_s and μ_s. When the cylinders are symmetrically displaced, R_l is inclined at ϕ_d but needs to be inclined at the larger ϕ_s to reach limiting statical friction; this occurs after some more movement of the right cylinder, Fig. 1.9(b.i).

As soon as the left cylinder can slip, R_l immediately reverts to the smaller inclination, ϕ_d, causing the intersection point to move up slightly. Since the right cylinder is closer to the rod centroid, the inclination of R_r drops below ϕ_d and, thus, below the limiting value altogether. The right cylinder stops moving, Fig. 1.9(b.ii), and we have, in effect, reversed the initial arrangement of slippage between the sides.

This state of motion continues until the left cylinder is sufficiently closer to the middle for R_r to become inclined again at ϕ_s and to start slipping; the left cylinder stops, and so forth until the cylinders meet close to the middle. This example can be easily demonstrated using a long ruler placed on two index fingers.

1.5   Final Remarks

Friction cannot be ignored even if it features minimally. Its constitutive behaviour is an inequality relationship between the responsible force components, which is satisfied only when slippage occurs. A 'safe' solution accords both equilibrium outcomes, of slippage or none, where the inequality has to be evaluated; this is over-wrought.

Pre-supposing slippage specifies the complexion of forces exactly, making geometrical or graphical solutions more amenable, especially when dealing with prismatic sections. The 'point mass' viewpoint of equilibrium so readily applied to, say, a block resting on a slope, is only applicable if moment equilibrium does not matter – even then, this requires a robust judgment. The actual geometry – of the body shape, support/contact conditions, and where the forces are applied – must be regarded for accuracy; their interaction equilibrium-wise, however, can become more varied, introducing other possible kinematical outcomes such as toppling and rolling.

2 Equilibrium of Pseudo-Deformed Bodies

Often loads applied to a structure do not depend on the deflections they induce; think of a gravitational body force or a contacting pressure between rigid bodies. If there is a dependency, they couple the very forces (and moments) in equilibrium to the displacements they induce, leading to *statical indeterminacy*.

In order to solve such problems, we must assume a displacement profile dictated by the *constitutive i.e.* characteristic behaviour of the loads they couple to. Its salient values remain to be found, but crucially these, and all other displacements, are relatively small compared to the overall scale of the problem. The loads may be actually very small, but such purpose allows us to express them accurately in terms of them whilst considering equilibrium in the undeformed state. This (mild) contrast in how we treat the role of displacements leads to a class of *pseudo-deformed* problems.

2.1 Foundational Loads

Figure 2.1(a.i) shows a rigid horizontal beam supported on elastic springs at both ends. This model provides a basic description of, say, how a stiff slab might settle vertically on 'softer' ground represented by the springs. A vertical force is applied a distance x from one end, and we wish to find the displaced shape. The linear stiffness of both springs is k, and we measure the vertical displacement δ at P.

An eccentric force insists that the beam also rotates in plane, by an angle θ to the horizontal. This is small enough that displacements arising from it are purely vertical, giving an offset linear profile throughout, Fig. 2.1(a.ii). The end displacements are 'absorbed' by each spring compressing, so we focus on their expressions: $e_1 = \delta - x \sin\theta$ and $e_2 = \delta + (L - x)\sin\theta$.

These are also small compared to L, and we can divide both by L and compare right-side terms. Clearly δ/L should be small, as should $\sin\theta$ compared to x/L, returning $\sin\theta \approx \theta$. As a result, $e_1 \approx \delta - x\theta$ and $e_2 \approx \delta + (L - x)\theta$.

The compressive forces in the springs push back against the beam to give end reactions, ke_1 and ke_2. A free-body diagram of the beam in its original level state, Fig. 2.1(b), enables us to write vertical force and moment equilibrium statements as:

$$P = ke_1 + ke_2, \quad x \cdot P = L \cdot ke_2. \tag{2.1}$$

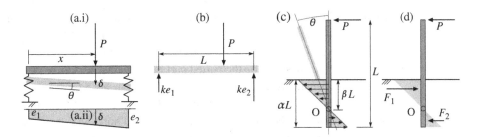

Figure 2.1 (a.i) Rigid beam supported on elastic end springs and deflecting under P; (a.ii) displacement profile with end values e_1 and e_2. (b) Forces applied to beam alone w.r.t. the initial configuration. (c) Partially buried rigid pole carrying a top load P, with horizontal ground displacements being linear about the centre of rotation O. (d) Equivalent underground forces, F_1 and F_2, applied to the pole at the centroids of the triangular pressure forces.

Substituting for e_1 and e_2 into the first statement, we find θ explicitly as $(P/k - 2\delta)/(L - 2x)$ which, when substituted into the second, returns $P/\delta = k/[2(x/L)^2 - 2(x/L) + 1]$.

Remembering that x is a singular location, the right-hand side and thus P/δ are constant. This ratio measures the *structural* stiffness, from the rate of change of applied force with the displacement of its point of application. Linearity between P and δ follows directly from the small displacement assumption where e_1 and e_2 themselves depend linearly on δ and θ.

When displacements are larger and all forces remain vertical, the same expression for P/δ is obtained when x is measured along the beam rather than horizontally because, and peculiar to this problem, the expression for moment equilibrium is unchanged.

A more realistic model of a compliant ground response is shown in Fig. 2.1(c). A vertical pole of length L is buried to a depth of αL before a horizontal force P is applied to its top end. The pole tends to rotate 'through' the ground, causing it to bear normally upon the buried depth in a linear elastic fashion.

In particular, the local elemental force, or *intensity* of loading, applied to a point on the buried depth is equal to the product of its lateral displacement and some constant k, equivalent to the elastic stiffness of the ground – altogether of units N/m. Displacements are again small so that equilibrium is assessed for the original vertical pole.

Unlike the previous example, we do not focus initially on the displacement of P: with θ, each spring displacement was conveniently expressed before eliminating the rotation from the final expression for the applied force. The system has, in fact, two *degrees-of-freedom* in δ and θ, which is also true presently, for the pole translates and rotates.

These freedoms are expressed alternatively by specifying a fixed point of rotation for the pole at some unknown location; it has to be located on the pole axis otherwise there will be a net movement of the entire pole vertically. Note the similarities to the rod problem with distributed friction in Chapter 1.

Denoting this point by O and setting it to be a distance βL below the surface, the two degrees-of-freedom are thus β and the small rotation θ. The value of β follows from satisfying force and moment equilibrium, found shortly: if it is larger than α or is negative, O then lies outside the buried part but remains valid.

Around O, small displacements of the buried pole are horizontal and build up linearly from zero, with maximum values of $\beta L\theta$ and $(\alpha - \beta)L\theta$, respectively. Multiplying by k, the distributed loading is also linear and horizontal, normal to the pole.

The total horizontal force from the ground can be found by integrating the loading intensity along the buried depth. This integration is also equal to the area enclosed by distribution, which can be sub-divided into a pair of triangles about O. The resultant *equivalent* force from each area is now easier to compute, with a line of action immediately located at the triangular centroid, which also simplifies moment equilibrium.

Denoting them by F_1 and F_2 in Fig. 2.1(d), $F_1 = (\beta L\theta/2) \cdot \beta L \cdot k = \beta^2 L^2 \theta k/2$ (top part, rightward) and is located $\beta L/3$ below the ground; $F_2 = (\alpha - \beta)^2 L^2 \theta k/2$ (bottom, leftward) at $(\alpha - \beta)L/3$ up from the bottom of the pole.

Taking moments about the top of the pole, we exclude P deliberately, in order to find the location of O directly:

$$F_1\left[(1 - \alpha)L + \frac{\beta L}{3}\right] - F_2\left[L - \frac{(\alpha - \beta)L}{3}\right] = 0. \tag{2.2}$$

Substituting for F_1 and F_2, expanding, and then cancelling common terms in L, k, and θ, we find:

$$\alpha^3 - 3\alpha^2 - 3\alpha^2\beta + 6\alpha\beta = 0 \quad \rightarrow \quad \beta = \frac{\alpha(3 - \alpha)}{6 - 3\alpha}. \tag{2.3}$$

For a specified value of α between zero and unity, β is always smaller and positive, i.e. O is always located below ground on the pole. The largest value of β is 2/3 when $\alpha = 1$, despite the pole being just fully buried.

F_1 and F_2, with β now replaced, oblige horizontal force equilibrium via $P - F_1 + F_2 = 0$, returning $P = \theta \cdot (\alpha^3 L^2 k/2)/(6 - 3\alpha)$: every term on the right-hand side is a specified constant except for θ, which expresses a linear relationship to P.

Finally, the horizontal displacement of P is θ times the distance from O: $L(1 - \alpha + \beta)$. If we wish, we can express the structural stiffness of the system by dividing P by this value.

2.2   Upthrust: Punting Safely

Other real loads that induce and depend on displacements stem from bodies immersed in fluids. In particular, Archimedes' principle tells us that the force, or *upthrust*, exerted by a fluid upwards on an object fully immersed or partially submerged is equal to the weight of the displaced fluid and acts through the centroid of the displaced volume.

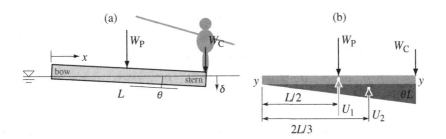

Figure 2.2 (a) Floating punt of uniform weight W_P being tilted by the weight of a chauffeur, W_C, at the stern. (b) External forces applied to the punt: U_1 and U_2 are separate forces from splitting the upthrust profile into simpler areas (volumes).

For example, the submerged depth of a uniform upright cube is a proportion of the overall height equal to the ratio of the cube and fluid densities. The *stability* of how this cube floats is treated momentarily.

First, consider the floating *punt* in Fig. 2.2(a). It is a low-lying, flat-bottom craft used nowadays to ferry tourist passengers around shallow tranquil town rivers. Propulsion is manually applied by an experienced chauffeur levering themselves against the river bed via a long pole. Optimal mechanical advantage comes from standing at the rear, or 'stern', of the punt, from where it is also easiest to steer using the pole as a rudder between exertions.

The tourists are 'evenly' distributed within the punt for simplicity, which descends uniformly to a height h above the surface before the chauffeur mounts rearwards. The downward forces are now the uniformly distributed self-weight of the punt and passengers, W_P, and that of the chauffeur, W_C, as shown. The punt clearly tilts in the water, becoming more submerged towards the rear. The flat bottom yields a linear depth profile which should not exceed h at the stern, otherwise water is taken on board.

The length of punt, L, is much larger than h, making the tilting rotation small but not negligible because the upthrust profile depends on the submerged shape; it is small enough, however, that the punt remains horizontal for a simplified equilibrium.

Adopting x as a horizontal coordinate from the bow (front) of the punt, the displacement depth profile, d, can be written as $d = y + x \tan \theta$, where y is the bow depth and θ is the tilt rotation. Alternatively, we could assume a point of rotation at some unknown location along the punt as per the previous buried pole example.

Whatever our approach, the upthrust is equal to the displaced volume times the fluid density, ρ, which is equal in turn to the submerged area in profile times ρ times a uniform (we assume) punt width, w.

But recognise that the submerged area is comprised of two prismatic sections, a rectangle, $y \times L$, and a triangle, $L \times \theta L$, assuming $\tan \theta \approx \theta$ for small angles, Fig. 2.2(b). The upthrust components from each are simpler expressions, straightforwardly located – as we did for the previous buried pole; halfway along for the rectangular part and two-thirds for the triangle.

Denoting these forces by U, we have $U_1 = yLw \cdot \rho$ and $U_2 = (\theta L^2/2)w \cdot \rho$. The final force diagram in Fig. 2.2(b) gives force and moment equilibrium (about $x = 0$) for a chauffeur at the very stern ($x = L$) as:

$$U_1 + U_2 = W_P + W_C, \quad U_1 \cdot (L/2) + U_2 \cdot (2L/3) = W_C L + W_P \cdot (L/2)$$

$$\rightarrow \quad U_1 = W_P - 2W_C, \quad U_2 = 3W_C. \tag{2.4}$$

From the definitions of U_1 and U_2 and the final two expressions above, we find y and θ explicitly to ensure an ultimate stern depth smaller than h:

$$h > \underbrace{\frac{W_P - 2W_C}{Lw\rho}}_{y} + \underbrace{\frac{3W_C}{(L^2/2)w\rho}}_{\theta} \cdot L \quad \rightarrow \quad hLw\rho > W_P + 4W_C, \tag{2.5}$$

where the final inequality can re-stated as $W_C < (hLw\rho - W_P)/4$.

The quantity, $hLw\rho$, is the upthrust acting on the punt and passengers alone when it is just immersed by some (unknown) agency, which must be larger than their combined weight: otherwise they would sink. The quantity on the right-hand side is therefore positive and tells us about the proportions (and mass) of the punt (and its passengers) compared to the chauffeur, to avoid sinking.

2.3   Float or Fall Over?

We finish with a floating uniform cuboid in Fig. 2.3(a), of unit length and general height b and width a. When the ratio of the cube density to that of the fluid is equal to α, the *specific* density, the submerged depth is αb when perfectly upright. Note that α is limited to unity before the cuboid sinks.

The cuboid will bob up and down if it is moved vertically and released, and eventually settle. On the other hand, if the cuboid is tilted – from sideways wind or wave loading, we might ask if it self-rights or keels over? Experience suggests that if the cuboid is more than half submerged with its centre of mass, G, below the water line, it will overturn if slender in height: a stick or a pen always floats horizontally. When the cuboid sits more out of the water than in it, G lies above the water line with a greater propensity towards toppling, we surmise.

A displaced configuration is again essential for calculating the correct upthrust. The level of movement does not have to be appreciable when considering stability, for any tendency to move away from upright will happen for any perturbation, no matter how small. We also note that the cuboid rotates about the mid-point of the original surface line where the now emerging and submerging parts, represented by the triangular portions in Fig. 2.3(b), are equal. This affords no change in the upthrust, and equilibrium of vertical forces remains assured.

For moment behaviour, consider first the tilted geometry in Fig. 2.3(c), which assumes that G originally lies below surface, *i.e.* $\alpha > 1/2$. We have divided the submerged part into two; a rectangle, ABCD, and a triangle, CDE. The area of ABCD is $a \cdot (\alpha b - a\theta/2)$ given that the right-side vertical edge in Fig. 2.3(b) rises $a\theta/2$ above

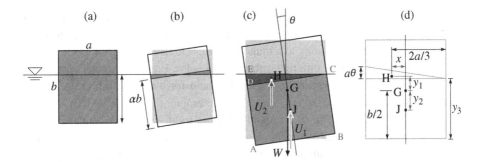

Figure 2.3 (a) Uniform cuboid ($a \times b \times 1$) floating in still water at a depth αb. (b) Cuboid rotates by a small angle θ; the darker grey triangles are equal in size. (c) Separate upthrust forces, U_1 and U_2 from the submerged areas (volumes) of the rectangle ABCD and triangle CDE, respectively. G, J, and H are centroids. (d) Detailed geometry for points from (c).

the surface – for θ small, of course. The triangular area is $a \cdot a\theta/2$, and each upthrust force, U_1 and U_2, multiplies these areas by ρ, the density of the fluid (recall that the cube has unit length into the page, equating its volume and area of cross-section). Their respective centroids are marked by points J and H as shown.

The cuboid tends to become upright because there is a net clockwise moment against the direction of θ. Three intrinsic coordinates are highlighted in Fig. 2.3(d) for a righted cuboid to aid this calculation: point H is separated from G by x and y_1 parallel to the width and height, and point J from G by height y_2. The upthrust components parallel to the height multiply U by $\cos \theta$, and to the width, $U \sin \theta$. A positive restoring moment about G is therefore given by

$$x \cdot U_2 \cos \theta + y_1 \cdot U_2 \sin \theta - y_2 \cdot U_1 \sin \theta \geq 0$$
$$\rightarrow \quad (x + y_1\theta)U_2 - y_2\theta U_1 \geq 0. \tag{2.6}$$

The geometry terms are straightforward to find: x is the centroidal span of the triangle minus the cuboid half-width, *i.e.* $2a/3 - a/2 = a/6$; y_1 is equal to the centroidal height of the triangle plus that of the triangle above G: $a\theta/3 + y_3 - b/2$ where, for compactness, $\alpha b - a\theta/2$ is written as y_3; and y_2 is $b/2 - y_3/2$. Substituting these into Eq. (2.6) along with U_1 and U_2, we obtain

$$\rho\frac{a^2\theta}{2}\left[\frac{a}{6} + \theta\left(-\frac{a\theta}{6} + \alpha b - \frac{b}{2}\right)\right] - \rho a\left(\alpha b - \frac{a\theta}{2}\right)\theta\left[\frac{b}{2} - \frac{\alpha b}{2} + \frac{a\theta}{4}\right] \geq 0. \tag{2.7}$$

On first sight this seems formidable, but θ, at least, is present in each term when we multiply out. Second-order terms, and higher, are much smaller, so they can be neglected in comparison. After some tidying up and remembering that ρ and θ are non-zero, we have from inside the brackets:

$$\rho\theta\left[\frac{a^3}{12} - \frac{ab^2\alpha}{2}(1 - \alpha)\right] \geq 0 \quad \rightarrow \quad \frac{a}{b} \geq \sqrt{6\alpha(1 - \alpha)}. \tag{2.8}$$

If we repeat the analysis with G initially above the surface so that $\alpha \leq 1/2$, we find the same expression, which now applies for the complete range of α between zero and unity, corresponding to a range from light to heavy materials.

To compare different cuboids, we solve Eq. (2.8) as the equality for b, which becomes the maximum height of cuboid for a fixed width at the point of overturning: $b_{max} = a/\sqrt{6\alpha(1-\alpha)}$. As an example, b_{max} is equal to $a/\sqrt{3/2} = 0.82a$ for $\alpha = 1/2$, giving us a cuboid which is mildly squat. By making the cuboid out of heavier material, α moves closer to unity and b_{max} becomes larger than a: in the limit as $\alpha \to 1$, the cuboid is very slender and mostly submerged; this we expect.

The performance is symmetrical about $\alpha = 1/2$ from Eq. (2.8), where b_{max} rises as α declines. This behaviour seems less obvious because it says that a lighter cuboid can become taller the more it sits out of the water before toppling. We should, nevertheless, expect this to be possible, and we can imagine, for example, a narrow polystyrene block floating mostly on top of water, behaving this way.

Splitting the upthrust into components from prismatic profiles, again, simplifies calculating their size and location, and ultimately the restoring moment. We can perform the same assessment by retaining a single upthrust force for the submerged compound shape; there are now two forces, where the upthrust line of action must lie outside of the cuboid's weight for a self-righting moment. The point where this line intersects an original vertical line in the cube through G is more commonly known as the *metacentre*. If this point lies above G, the cuboid is stable, and *vice versa*; we do not have to calculate forces – rather, only their geometry.

On first sight, this seems to be less involved because we are only dealing with geometry. But it is equivalent in effort to our method because the centroid of the single upthrust is calculated by using the same prismatic shapes: the reader should verify that Eq. (2.8) can be identically found this way.

Finally, we note that the cube will invariably overshoot the vertical when it self-rights. The geometry in Fig. 2.3(c) is now reflected about the vertical, which preserves the restoring tendency. The cube will eventually settle after some oscillation – damped, of course, by the fluid around it.

2.4   Final Remarks

The equilibrium solution to these problems is defined by the very displacements the applied forces induce, but we escape an exact analysis using the actual geometry by insisting on small displacements. Only rigid-body movements have been accorded, without the complexity of elastic deformation; later, in Chapter 20, we consider how contacting force profiles change when elasticity prevails.

When the applied forces result in a distributed reaction applied to the body, we have broken them down into simpler components defined by the 'geometry' of their distribution: their net sizes and their points of application become more straightforward to use, and the analysis is more transparent.

A floating cuboid is stable when buoyancy forces are 'restoring', similar to the action of an extended spring upon a connected mass. No matter the spring force, the mass is conveyed to zero extension – at least on the horizontal plane. The limit of zero restoration is therefore a spring with zero stiffness; conversely, and peculiarly, an 'unstable' spring would have negative stiffness. It is precisely this crossover in stiffness, in mathematical terms, which predicates general instability in structures.

3 Simple Cables

The simplest loaded cable is straight and vertical with a weight attached at one end. The other end may pass over a roller or cylinder as part of a larger structure to be connected elsewhere. A free-standing cable connected to fixed, pinned supports can also carry transverse loads, again by transmitting tension along its length.

The simplest case is a point force applied transversely to a straight cable, which divides into two straight, inclined parts about the loading point, as shown in Fig. 3.1(a). The discontinuity in the cable gradient is essential for equilibrium of the tensions on either side, but the cable now stretches in length, and elastically, we assume.

The equilibrium response to a distributed loading is more involved but is achieved by the tension varying along the cable, which is no longer piece-wise straight. An element of the loaded cable is shown in Fig. 3.1(b) within a Cartesian (x, y) coordinate system. The transverse (vertical) direction of loading intensity w is y and an element of cable of horizontal width δx bears a force $w\delta x$.

Between the ends, separated in height by δy, there are elemental variations in tension from T to $T + \delta T$ and in horizontal inclination from θ to $\theta + \delta\theta$. Horizontal and vertical components of tension are shown as H and V, but we first resolve vertically in terms of T:

$$w\delta x + T \sin\theta - (T + \delta T)\sin(\theta + \delta\theta) = 0$$
$$\rightarrow \quad (T + \delta T)(\sin\theta \cos\delta\theta + \cos\theta \sin\delta\theta) - T \sin\theta = w\delta x. \qquad (3.1)$$

The term $\sin\delta\theta$ is approximately $\delta\theta$ and $\cos\delta\theta \approx 1$. Infinitesimal product terms can be ignored after multiplying out, and dividing by δx before observing the limit we find:

$$\frac{d}{dx}(T \sin\theta) = w. \qquad (3.2)$$

This is also equivalent to $dV/dx = w$, a more obvious statement from Fig. 3.1(b). In the horizontal direction we have equivalently $d(T \cos\theta)/dx = 0$, setting the horizontal component of tension, H, to be constant – only because there is no loading intensity in that direction. The gradient, dy/dx, equal to $\tan\theta$ and thence, V/H, provides a substitution for y:

$$\frac{dV}{dx} = w \quad \rightarrow \quad \frac{d}{dx}\left(H\frac{dy}{dx}\right) = w \quad \rightarrow \quad \frac{d^2y}{dx^2} = \frac{w}{H}. \qquad (3.3)$$

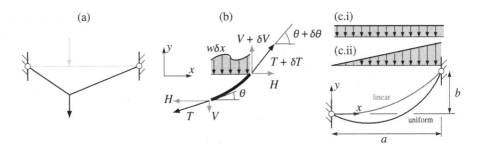

Figure 3.1 (a) Point-loaded cable with straight parts either side. (b) Equilibrium performance of an element of cable carrying a loading intensity w; the tension T can be resolved into horizontal and vertical components, H and V. (c) Loaded cable slung between asymmetrical supports; (c.i) uniform loading and (c.ii) linear loading.

Note that taking moments merely confirms the earlier gradient statement between V and H, and no further information is gleaned.

We have a single equation governing the transverse displacements, y, in terms of w and H. It mirrors the governing equation of bending moment equilibrium for an elastic beam introduced later where, equivalently, w/H would be the loading intensity and y the bending moment; this permits some interesting comparisons in Chapter 7. When cable gradients are shallow, d^2y/dx^2 corresponds to the displacement curvature (*c.f.* 'Author's note').

Equation (3.3) can be integrated twice with respect to x to find y provided the two constants of integration can be related to the cable shape. However, if w is specified as a polynomial expression in x, the shape described by y must be two *orders* greater according to Eq. (3.3). Our expectations of shape are thus girded by a general polynomial in y whose coefficients can be calculated from known features of the cable.

3.1 An Example

The cable in Fig. 3.1(c) is slung between two (pinned) supports separated in width and height by a and b as shown. In the first case, there is a uniform loading intensity akin to self-weight, (c.i); in the second, the loading is linear, being zero at one end where the cable is also horizontal, (c.ii).

The first cable shape is a quadratic function in x because the constant load is obviously of order zero when written as $w \times x^0$. Specifying the (x, y) origin to be at the left support, $y = Ax^2 + Bx$ without an intercept because of where the origin is: A and B are unknown coefficients. From the position of the right-side support, $(x, y) = (a, b)$, we have $b = Aa^2 + Ba$, and differentiating y twice we find $A = w/2H$.

The coefficients are now known in terms of a, b and H. In this regard, H is also the horizontal reaction at each support, and its value specifies the cable shape absolutely. Conversely, if extra information about the shape is specified, such as a certain gradient, H will be unique.

In the second case, we may write $y = Ax^3 + Bx^2 + Cx$ for the same coordinate origin. Zero gradient at $x = 0$ sets C to be zero; furthermore, with zero loading intensity at the same place, B is zero after differentiating y twice. From the right side, $b = A\,a^3$, which sets A and the final shape, $y = x^3 \cdot (b/a^3)$. This does not depend on H because the loading requirements at $x = 0$ are sufficient to specify the problem completely. In particular, if the amplitude of w is W at $x = a$, $d^2y/dx^2 = W(x/a)/H = 6bx/a^2$, which sets the relationship between W and H.

When only point forces are applied to a light cable, $w = 0$ and the *complementary function* of Eq. (3.3) reveals a linear variation in y, which are the inclined, straight cable portions between the loading points we expect. Often, transverse cable displacements are larger than axial extensions, and for a relative displacement δ over a cable portion of original length L, its new length is obviously $\sqrt{(L^2 + \delta^2)}$.

The strain compares the change in length to the original, which can be shown to be $\sqrt{1 + (\delta/L)^2} - 1$. Relative displacements, however, remain small and δ/L is much less than unity, where the Binomial Theorem sets $\sqrt{1 + (\delta/L)^2} \approx 1 + (1/2) \cdot (\delta/L)^2$ and a strain equal to $(1/2) \cdot (\delta/L)^2$.

This is very small indeed given the squared dependency on an already small term; but it is not negligible, otherwise there is no stress or tension after multiplying by the Young's Modulus, E, and its cross-sectional area, A. These resultants therefore depend on $(\delta/L)^2$ also.

In traditional statics problems, there is usually a linear dependency between the loads, displacements and internal stresses, but we have just found a geometrically nonlinear relationship. The rate of change of tension with displacement also tells us informally about stiffness, which must go with δ above. However, when δ equals zero – just as the point loading is applied, there is paradoxically *no* stiffness.

Both of these features emerge from the particulars of this problem, of a point force applied normally to a previously stress-free cable. If the cable does not deflect, it is not possible for the load to exert components of cable tension, so there is no resistance and no stiffness. The latter only builds up with deflection, specifically from sufficient changes in geometry. It is possible to impart initial stiffness by *pre-stressing* the cable before loading; we deal with a similar case for bars in Chapter 4.

To recap: point forces produce linear displacements whilst distributed loadings are equilibrated by a displaced shape two orders greater than their polynomial order – if defined this way. We (informally) understood strain in the sense of the *change* in geometry between the unloaded and loaded states for a given point force. Equilibrium is also satisfied for this loading if we assemble longer but *rigid* cable portions in the same displaced configuration. In other words, Eq. (3.3) is purely an equilibrium statement based on the current shape from y for any loading.

Calculating the exact strain depends on how far we move from the initial shape, but even then we allow only reasonable changes in geometry for a tractable analysis. If, for example, the cable is originally horizontal, a strained element has a new length, $ds = \sqrt{(dx^2 + dy^2)}$. This gives a local strain equal to $(ds - dx)/dx$, which reduces to

$$\frac{\sqrt{(dx^2 + dy^2)}}{dx} - 1 = \sqrt{1 + \left(\frac{dy}{dx}\right)^2} - 1 \approx \frac{1}{2}\left(\frac{dy}{dx}\right)^2 \tag{3.4}$$

when the gradient, dy/dx, can be safely assumed to be much less than unity. Anything larger demands a more sophisticated analysis beyond our scope.

3.2 Cables: Smooth and Rough

In practice, a point force may be applied to a cable through bearing of a relatively small rigid cylinder, as shown in Fig. 3.2(a). A smooth cylinder, we know, allows for slippage without resistance, where the components of cable tension on either side are always assumed to be equal. But the cable subtends a given angle on the surface of the cylinder, which dictates the relative direction of tensions on either side, and these directions must not only satisfy local equilibrium with the force itself but also fit with the overall deformed geometry of the cable.

The gently inclined, elastic cable in Fig. 3.2(b) indicates what we mean. It is attached to supports separated in width by x and a very small height, y, in comparison. A deadweight, P, is applied quasi-statically to the cable via a small but smooth loading cylinder and finally settles off-centre, as shown; it must lie below the lowest support to avoid trundling towards it. E and A are the cable's Young's Modulus and area, respectively.

Either side of P, the displaced cable portions are both inclined at the same angle θ for equal tensions, and have lengths a and b. Neglecting the size of the cylinder sets $y = b \sin \theta - a \sin \theta$ and $x = (a + b) \cos \theta$, from which we find: $a = (x/\cos \theta - y/\sin \theta)/2$ and $b = (x/\cos \theta + y/\sin \theta)/2$. The overall displaced length, $a + b$, is now equal to $x/\cos \theta$. The original cable length is approximately x having stipulated y/x to be much less than unity, giving a cable extension equal to $x/\cos \theta - x$.

The cable tensions are both equal to $(P/2) \sin \theta$ from equilibrium. Dividing by EA, we obtain the cable strain, which must equal $(x/\cos \theta - x)/x$. Thus, we have a unique relationship between the applied force and θ, which dictates the equilibrium geometry, *i.e.* $P = 2EA (\tan \theta - \sin \theta)$.

We can solve this expression for θ in terms of P but it is easier the other way round; stipulating θ sets P given EA (or P/EA), and then a and b – how the deformed cable looks. Our informal solution contends with statical indeterminacy *and* geometrical nonlinearity, made much easier here by the gentle incline of the cable.

We can examine whether 'smooth' cable tensions are, in fact, equal by assuming a rough cylinder and unequal tensions and examining what happens when the corresponding coefficient of friction, μ, becomes zero. In general, the cable tension increases from T_1 to T_2 over its contacting length, which subtends a total angle (this time) of θ on the cylinder, as shown in Fig. 3.2(c).

Frictional forces are distributed over θ and oppose the increase in tension, suggesting cable slippage in the general direction of T_2. Note that the other external forces required for equilibrium of the cylinder are not shown and the ability of the cable to stretch elastically is neglected, in order for frictional contact to be treated most simply.

A locally contacting element of cable subtending $\delta\theta$ carries a differential tension δT and a normal reaction force δN. The corresponding friction force, δF, is limiting

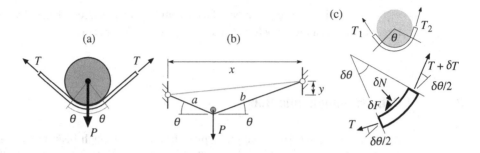

Figure 3.2 (a) Smooth bearing of a cylinder on a cable portion with equal tensions either side. (b) Equilibrium shape of a displaced asymmetrical cable where the point loading is transmitted from bearing of smooth cylinder. (c) Frictional contact between a cylinder and a wrapped cable leading to a change in tension across the cylinder.

here and everywhere between the cable and the cylinder. Resolving normally and tangentially to the element we find

$$\delta T \cos(\delta\theta/2) - \delta F = 0, \quad [(T + \delta T) + T]\sin(\delta\theta/2) = \delta N$$

$$\rightarrow \quad \delta T = \delta F, \quad T\delta\theta = \delta N. \tag{3.5}$$

Replacing δF with $\mu \delta N$, equating both expressions and observing the limit yields $dT/T = \mu d\theta$. Integrating both sides:

$$\log(T_2/T_1) = \mu\theta \quad \rightarrow \quad \frac{T_2}{T_1} = \exp(\mu\theta). \tag{3.6}$$

If T_2 happens to be smaller than T_1, slippage reverses and we invert their ratio above. Importantly, the tensions are equal only when μ is zero.

As an example of the 'variety' afforded by the presence of friction, Fig. 3.3(a) shows a light cable symmetrically draped over a pair of fixed, rough cylinders with equal portions overhanging vertically. Weights W hang from each end, and a central vertical point force, P, equal to a multiple α of the total load, $P = 2\alpha W$, is applied.

The cable portions on the inside are now both inclined at θ to the horizontal, and slippage inwards or outwards, relative to the centre, is feasible with P or the weights moving downwards (and the weights or P moving upwards). Given α, we seek to characterise this behaviour in terms of the cable shape (via θ) and the coefficient of friction. The position of the cylinders can be initially set so that all configurations from $\theta = 0°$ to $90°$ are possible for a finite length of cable. A symmetrical layout is presumed only for simplicity; the middle cylinder, therefore, can be smooth or rough with equal tensions either side.

These tensions are both $(P/2)\sin\theta$. Over each outer cylinder, the cable wraps an angle $\pi/2+\theta$, and the tension ratio across them is equal to $(P/2\sin\theta)/W = \alpha/\sin\theta$. The cable is about to slip inwards when this ratio is greater than unity and equal to $\exp[\mu(\pi/2 + \theta)]$ from Eq. (3.6); if the cable slips outwards with the weights moving

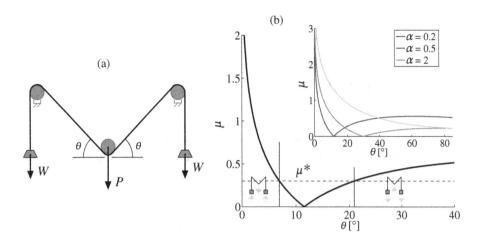

Figure 3.3 (a) Cable draped symmetrically over a pair of rough, fixed outer cylinders and connected to weights W; equilibrium is maintained by a central vertical force P. (b) Limiting coefficient of friction, μ, in terms of cable inclination, θ, and load ratio α $(= P/2W)$; a given coefficient is μ^* and the inset figure accords different load ratios. Configurations with equal tensions occur when $\mu = 0$.

down, we invert the tension ratio. The final dual expression is, therefore, $\alpha/\sin\theta = \exp[\pm\mu(\pi/2+\theta)]$.

Figure 3.3(b) indicates the variation in μ with θ for P much lighter than the total load, *e.g.* $\alpha = 1/5$. At $\theta \approx 11.5°$, $\mu = 0$ and the system is *balanced* with equal tensions across the cylinders. This, however, is a very precise requirement because μ is otherwise non-zero.

For smaller values of θ, the required μ has to increase rapidly; even though the applied load is fixed, it foists a larger tension as the cable flattens, causing the tension ratio across each cylinder to rise sharply. Lying below this part of the curve, we have insufficient friction, and the cable slips inwards. For larger values of θ, the cable tension reduces, becoming smaller compared to that, weight-side. Slippage is now outwards with the threshold for μ rising somewhat more gently.

In practice, μ is fixed, giving us a horizontal line, $\mu = \mu^*$, drawn on Fig. 3.3(b). Depending on the value of μ^*, we may see one or two intersection points corresponding to limiting equilibrium geometries; here $\mu^* = 0.3$, giving two such points.

Just before the first point where $\theta \approx 7°$, the configuration will slip inwards; afterwards, it remains static up to the second point ($\theta \approx 21°$), where beyond it will slip outwards. This provides a simple, practical way for determining μ, which can be doubly confirmed in value if two intersection points prevail; assurance of the latter is given by adjusting the ratio of weights to force P through α.

The performance for three different values of α is given in Fig. 3.3(b). Again, with $\alpha < 1$, we see slippage both inwards and outwards; when $\alpha = 2$, the cable can only slip inwards. The transition to this mode occurs when μ is zero and $\theta = \pi/2$, which sets $\log(\alpha/\sin\theta) = 0$, that is $\alpha = \sin\theta = 1$. We may conclude that if P

is smaller than the total weight, there are two possible slipping modes, otherwise the arrangement falls inwards.

3.3   Final Remarks

Equation (3.3) is about equilibrium of the *current* shape of the cable. The initial shape does not feature except for its *boundary conditions* – how it is connected to the ground. In this sense, a flexible cable *rigid* in axial extensions can be a viable solution for a given loading, provided its length exactly fits the support geometry.

A cable under uniform self-weight alone has a quadratic profile. Its displacements and gradients are sufficiently small that the loading is not altered by the displacements it causes. If they are not small, however, we must assume that its self-weight is not evenly spread but becomes more concentrated as the gradient increases. The adjustment to the parabolic shape can be found from the same elemental treatment in Fig. 3.1 where the applied loading is calculated from the true arc-length, ds, *i.e.* $w\delta s$ rather than $w\delta x$.

Consequently, Eq. (3.2) and those that follow are expressed with respect to ds instead. Solving them within a Cartesian framework is desired because this describes the workings of the geometry, its span and central deflection (or dip). A change of variable is thus employed, leading to a more involved integration. We do not pursue a formal solution other than to note that one final expression is $y = (H/w) \cdot [\cosh(wx/H) - 1]$ when the origin is located at the dip: *c.f.* the small displacement solution, $y = wx^2/2H$. The former shape is more commonly known as a *catenary* (Latin: 'chain').

The reader can compare these two expressions for different ratios of w/H, where a larger ratio is equivalent to a heavier cable and higher displacements, and *vice versa*. For a value of $4/5$, the central displacement is equal to one fifth of the span, which, ordinarily, would be considered relatively large.

Comparing the displacement profiles, however, the largest difference, at the ends (because the origin of both is at the dip), is equal to only 1.3 %. The catenary refinement of behaviour is wholly accurate for sizeable displacements, but the quadratic description is not far removed; the simplest of models can often suffice.

Part II

Truss Frameworks

4   Statical (In)determinacy in Trusses

Put simply, Maxwell's original rule is foremost a mathematical statement of the rigidity requirement for a truss framework of pin-jointed bars. Rigidity implies absolute stiffness with no possible displacements, otherwise the framework collapses as a mechanism. Later, this rule was modified[1] to account explicitly for mechanisms as well as states of internal 'self-stress'. In certain cases, these can co-exist with profound implications for practical structural behaviour.

4.1   The Rule

A truss is generally connected to the ground by support joints that prevent displacements in one direction at least; external forces may be applied to any, or all of, the remaining joints. The unknown statical quantities are thus the tensions in b bars along with r reaction *components*. From j joints overall in D dimensions, being two or three, we have $D \cdot j$ equilibrium equations available.

Maxwell's Rule for these trusses says that $b + r$ must be at least equal to $D \cdot j$ for rigidity; and if $b + r > D \cdot j$, there are more bars or reactions than can be solved uniquely from equilibrium alone, giving us statical indeterminacy or *hyperstatic* behaviour. These surplus quantities are usually denoted as *redundancies*, not because they are inoperable but rather because the structure is stiff without them already.

Immediately, we can make substantive conclusions. First, a positive difference between $b + r$ and $D \cdot j$ indicates the number of redundancies; but only the *sum* of $b+r$ matters and not specific bars or reactions in view of how we confer redundancies. Conversely, we can choose any of them to be candidates for redundancies – in view of any further analysis. Second, the loading is not involved, and indeterminacy is wrought only by the layout of the framework and how it is connected to the ground. There is, of course, a consequence for how loads are eventually carried, successfully or not, but they do not cause indeterminacy.

The example of a simple symmetrical arch in Fig. 4.1(a) is statically determinate. Each support joint carries two reactions, $r = 4$, and two bars set $b = 2$; three joints overall in two dimensions gives $D \cdot j = 6$, which exactly balances $b + r$.

---

[1] C R Calladine, Buckminster Fuller's 'Tensegrity' structures and Clerk Maxwell's rules for the construction of stiff frames, *International Journal of Solids and Structures*, 14(2), pp. 161–172, 1978.

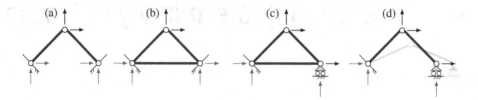

Figure 4.1 (a) Statically determinate pin-jointed arch: components of reaction forces (*r*) are grey arrows and applied loads, black. (b) Horizontal bar added between supports leads to indeterminacy. (c) Right support from (b) converted to roller support, to restore determinacy. (d) Horizontal bar from (c) removed, giving a mechanism.

Adding an extra horizontal bar between the supports, Fig. 4.1(b), increases *b* to three and nothing else. We now have a single redundancy, which can be the tension in the extra bar – or any other force quantity, as we please.

If we replace the right-side support with a roller pin, Fig. 4.1(c), *r* is reduced to three, bringing $b + r$ back to six again, which balances $D \cdot j$; by removing the possibility of a horizontal reaction force in the previous indeterminate case, we restore determinacy throughout.

Further removal, say, of the bar inserted in Fig. 4.1(b), clearly produces a mechanism in Fig. 4.1(d), which can fall flat when loaded: $b + r$ is equal to five, which is one less than $D \cdot j$.

Mathematically, decreasing *b* or *r* by one in a determinate case produces a mechanism: physically removing a bar or support constraint – for a structure which is basically rigid, is catastrophic. But if the framework is already indeterminate, the same action eliminates one redundancy, bringing us closer to determinacy.

Conversely, adding more bars or increasing *r* 'raises' the level of indeterminacy and redundancies. In order to calculate them, we must consider the *geometry* of deformation, specifically the connection between joint displacements and axial strains in the bars. This produces more unknowns, but even more relationships between them when we also consider the generalised Hooke's Law, giving us a completely soluble system: this is for later in Chapter 5.

But we ask the question what level of redundancy? Structures should not be mechanisms, but equally most practical structures are statically indeterminate. Having to do more analysis in order to find bar forces should not deter using them, but they offer two key improvements over determinate structures.

First, they are safer in an obvious way: if a bar fails or a support gives way, there is structural contingency in what remains. Less obviously, the same affords, paradoxically, analytical simplicity if we design them according to the Lower Bound Theorem, provided they can yield in a ductile manner: detailed elaboration is delivered in Chapter 6. Finally, they are usually stiffer than determinate structures and deflect less because some of the supports are overly constrained and more material is used than is needed.

Mechanisms should not, however, be dismissed out of hand. The vast bulk of *deployable*[2] structures rely on mechanistic actions for their opening or unfurling from a compactly stored configuration; from landing gear on aircraft to foldable solar panels on spacecraft, even to the average ironing board and umbrella. They do have to be stiffened after opening to become truly structural – in ways not dealt with here.

4.2   The Improved Rule

Our assessment thus far is one of counting based on bar connectivity. The specific geometry of pin-joint positions and supports *etc.* does not feature but obviously they do later during the actual computation of statical quantities. Counting and computation should, therefore, tally in the sense of their collective outcomes; if they do not, then we should examine why.

For example, our original statically determinate arch is now loaded centrally and vertically in Fig. 4.2(a). It is clearly equilibrated by equal bar tensions and the central joint displaces (a small amount) vertically for some bar elasticity. A shallower arch behaves the same, and we can imagine constructing it using the same supports but with shorter bars.

A completely flat arch, as shown in Fig. 4.2(b), behaves very differently. Horizontal bar tensions cannot equilibrate the applied vertical force, so both are zero. There is nothing to resist movement of the central joint, which *must* displace, causing the bars to rotate and to stretch before equilibrium is achieved vertically, Fig. 4.2(c). The arch must behave as a mechanism – if instantaneously in the context of displacements that follow.

Changing the geometry of the framework in Fig. 4.2(a) should not alter the fact that it *is* determinate: in moving from this state to a mechanism, we have not, as noted before, removed a bar or reaction component. We seemingly have a paradox.

A mechanism can be confirmed directly from equilibrium of the deformed configuration in Fig. 4.2(c) before examining what happens initially. The central joint displaces δ vertically where the now inclined bars both have an extended length $\sqrt{L^2 + \delta^2}$ approximately equal to $L[1 + (1/2) \cdot (\delta/L)^2]$ when δ is relatively small.

The corresponding bar strains match, unsurprisingly, the non-linear expression for displaced cable strains from Eq. (3.4), which is $(1/2) \cdot (\delta/L)^2$. Both tensions, T, are inclined horizontally at small angles δ/L so that vertically $P = 2T \sin(\delta/L) \approx 2T\delta/L$.

For equal cross-sectional areas, A, and Young's Modulus, E, the bar tensions are EA multiplied by the strain, setting P equal to $EA \cdot (\delta/L)^3$. The applied force is now a *cubic* function of joint displacement when ordinary truss behaviour is usually linear:

---

[2] K Miura and S Pellegrino, *Forms and Concepts for Lightweight Structures*, Chapters 8 and 9, Cambridge University Press, 2020.

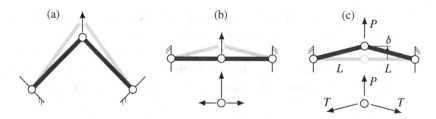

Figure 4.2 (a) Pin-jointed determinate arch. (b) The same arch as in (a) but originally flat and stress-free: however, bar tensions cannot be equilibrated for an applied load. (c) Displaced arch from (b) enables bar tensions to build up during loading.

the stiffness of the system is, therefore, given by the rate of change of P with respect to δ rather than dividing P by δ.

Differentiating accordingly, the stiffness is a function of δ^2; it therefore increases with the displacement but initially from *zero*. Thus, we have our original mechanism.

We can also revisit Maxwell's Rule for a conceptual interpretation. First, the number of redundancies is explicitly assigned as s, enabling us to write $s = b + r - D \cdot j$. When s is zero, the framework is statically determinate where removing a bar (or reaction), as we already know, produces a mechanism. But rather than finding $s = -1$ in our modified rule, we assign another, separate parameter for mechanistic consequence, m, which we subtract (so that m signifies positively the degree of mechanism – for we may remove more than one bar or reaction). The final modified rule can be written:

$$s - m = b + r - D \cdot j. \tag{4.1}$$

For the flat arch $s - m$ is zero with m equal to one. Consequently, s must also be unity somehow without changing the physical layout of the structure. A clue is provided in the original definition of s, of surplus statical quantities. Of course, adding bars or reactions increases s, but this is also implied by *extra*, if insoluble, equilibrium solutions for the *same* arrangement of bars.

To achieve this, consider making the flat arch in Fig. 4.3(a) by connecting together bars that are too short to meet naturally in the middle. A pair of external forces, T, must be applied to cause separate extensions in each bar until their ends overlap, where the connection can be made and the forces removed (these forces need not be equal to achieve overlapping), to give the same arrangement where $s - m = 0$.

Each bar, however, cannot shrink back, and horizontal equilibrium of the central joint sets their tensions to be equal. Moreover, without external loads explicitly applied, we have a state of *self*-equilibrating tensions, or self-stresses in general, and it is the word 'self' that enunciates s in the modified Maxwell Rule.

Infinitely many values of T satisfy equilibrium, but only one complies with the kinematics of how the bars are reconnected, *i.e.* geometrical compatibility. No other relationship exists between the bar tensions – they must be equal and thus represent a single solution set. This gives us $s = 1$, and we have resolved the paradox.

Note that we can only perform this thought-experiment with shortened bars; if they happen to be too long, they can be simply reconnected by rotating above the horizontal

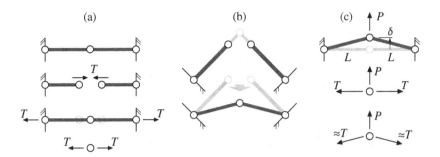

Figure 4.3 (a) Making a flat arch from bars that are too short initially. Each end joint has to be pulled towards the other to create the central pin-joint. (b) Bars that are too long initially always form a proper (inclined) arch without bar forces before loading. (c) Pre-stressed flat arch made as in (a).

line without exacting bar tensions. For the same reason, the original arch cannot attain a state of self-stress if its original bars are slightly too short before connecting them, see Fig. 4.3(b): it is clearly not a mechanism, and both s and m are zero for the same arrangement of bars and supports.

A flat arch constructed in the manner of Fig. 4.3(a) before loading is said to be *pre-stressed*. It is also immediately stiff upon loading, which can be verified again from its deformed equilibrium response. The previous kinematics apply where the non-linear strains of the inclined bars add to those from pre-stressing.

The size of the former, however, dominates the latter owing to its second order variation with $(\delta/L)^2$. Ignoring this contribution, the bar tensions are approximately constant and equal to their self-stress values, T. Vertical equilibrium is unchanged from $P = 2T\delta/L$, Fig. 4.3(c), giving an overall stiffness of $2T/L$ from the onset of loading. Without T, we again see mechanistic behaviour.

The modified Maxwell statement for the flat arch does not assert definitively that there has to be a mechanism or there is self-stress but that either is *possible*. Furthermore, the mechanism is declared to be *infinitesimal* because it is apparent only for negligibly small displacements of the central joint – it would not be very good as a deployable structure because of this limited freedom of motion. On the other hand, the final structure in Fig. 4.1(d) is clearly a continuous, or *finite*, mechanism with displacements unimpeded (up until the point where the central joint hits the floor).

4.3   Final Remarks

For more detailed truss layouts, self-stress and mechanisms may co-exist simultaneously and in a more obvious manner. In Fig. 4.4(a), the original tower is simply triangulated and statically determinate: $b = 9$, $j = 6$ and $r = 3$, thus $b + r = 12$, which is equal to $D \cdot j$. The layout is regular with angles of $0°$, $45°$ or $90°$, and all bars have length L or $\sqrt{2}L$.

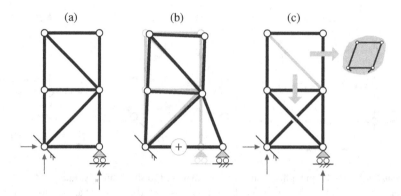

Figure 4.4 (a) Statically determinate truss tower with two 'bays'. (b) Bottom bar is elongated by some external agency, *e.g.* by heating it; the other bars in the bottom bay adjust their inclinations without straining, and the top bay also rotates. (c) Transposing the upper diagonal to the bottom bay produces a mechanism on top and an indeterminate bottom bay.

If the bar lengths change in some way, for example, from differential thermal expansion, the regular layout is compromised but, crucially, everything fits together without incurring bar forces. Each of the individual triangular 'bays' can be distorted, quite substantially in a rigid-body sense, provided the longest side is shorter than the other two added together: in Fig. 4.4(b), we show this when the bottom bar has an exaggerated extension (the source of which is not relevant in the sense of kinematics alone).

We now transpose the top diagonal bar to the bottom section and reconnect it, as shown in Fig. 4.4(c). Nothing has changed quantitatively, so $s - m$ remains zero. But the top section is clearly a mechanism – three-bar or four-bar, depending on your viewpoint – and certainly accords *finite*, not infinitesimal, joint displacements. In the bottom, the extra bar can be reconnected perfectly without inducing extra bar forces if its length remains as $\sqrt{2}L$, provided there are no loads.

Invariably, there will be self-stresses and, notwithstanding the upper section mechanistic behaviour, the bottom section will distort irregularly. The extra diagonal member will not fit perfectly, no matter how hard we try: we have to prise the end joints apart or force together for a compatible fit. Consequently, there is always a state of self-equilibrating tensions, giving $s = 1$ along with the obvious $m = 1$ on top.

In the next chapter (Chapter 5), we solve for bar forces in indeterminate trusses ($s > 0$) without any local mechanisms ($m = 0$) by the method of Virtual Work, expeditiously – by dealing thoughtfully with redundancies. Again, we shall see the contradictions of indeterminacy laid bare: that even though the structure is now more physically constrained than it has to be, there is greater solution freedom that we turn to our advantage.

5   Virtual Work for Trusses

The energy method of Virtual Work beguiles and empowers. Equilibrium and compatibility must only be satisfied *separately* when ensuring the balance of external work done and internal strain energy; this is the beguiling part, especially since one begets the other in reality. Separability, however, allows us to abstract the actual structural performance from any other arbitrary or *virtual* imposition, which, if expertly appointed, reveals our interested quantity directly: this is its power.

The imposition operates reciprocally in the sense of statical *vs* kinematical. For example, if we are interested in a displacement, we impose a virtual force in that direction: any magnitude will do, but often unity is simplest. For pin-jointed trusses, the method states:

$$\Sigma_{\text{joints}} W \cdot \Delta = \Sigma_{\text{bars}} T \cdot e. \tag{5.1}$$

The joint displacements, Δ, and bar extensions, e, are sufficiently small so that the bar tensions, T, and applied joint loads, W, do not change during work: the integral of effort is, thus, their respective dot-products. The bar tensions and extensions are aligned so we can take their simple product on the right-hand side and we often anticipate as such for the left.

All tensions and loads must be in equilibrium and can be defined independently of e and Δ, which must form a compatible set, synchronising perfectly in geometry terms without the bars mis-fitting or fracturing as they mete out the joint displacements.

When the structure is statically determinate, its virtual equilibrium response will be defined uniquely and the method executed ordinarily. Its efficacy is, however, enhanced for indeterminate structures because there are as many *extra* equilibrium responses as there are redundancies. There is more freedom, therefore, to 'short-cut' our virtual response towards the ultimate solution.

We now demonstrate this in the first two examples of indeterminate trusses. Before we can find joint displacements *etc.*, we must calculate the exact bar forces including the redundant bar tensions. For this, we choose virtual systems that are paradoxically unloaded, which sets the left-hand side of Eq. (5.1) to be zero from the outset, according a further economy of solution.

We return to a determinate truss in the third example to find a particular joint displacement by imposing more than one virtual loading. This approach counterpoints the usual fixation with only the joint in question. Our purpose is connected again to enabling the simplest virtual equilibrium response, which ultimately reduces our working even if the left-hand side of Eq. (5.1) has more terms than usual.

5.1   Example One

The square truss in Fig. 5.1(a) carries a diagonal load W at joint C. From Maxwell's Rule there is one redundancy, $b + r - D \cdot j = 6 + 3 - 2 \times 4 = 1 = s$, which we declare to be the unknown tension, R, in bar AC. We want to find R most efficiently, followed by the displacement of C in the direction of W. All bars have the same area of cross-section A, and Young's Modulus E, and are either L or $\sqrt{2}L$ in length.

We begin by determining as much actual, or *real*, information as possible, including the current bar tensions in terms of W and R. Equilibrium of joint C in Fig. 5.1(b) reveals T_{CD} and T_{CB} both equal to $(W - R)/\sqrt{2}$. From the other joints, the remaining peripheral bars have the same tension, and the other diagonal bar tension is $R - W$. Interestingly, the reaction at joint D is zero, and at A there is W equal and opposite to the applied loading.

The corresponding bar extensions stem from multiplying the bar tensions by the ratio of each bar length to AE (see Table 5.1). The real joint displacements are general, but only δ_C is defined explicitly in the direction of W at joint C, Fig. 5.1(a).

Since R already features in the real bar extensions, we can use another, virtual equilibrium set in order to calculate it. Denoting these quantities by a superscript $*$, Eq. (5.1) is first re-written as:

$$\Sigma_{joints} W^* \cdot \Delta = \Sigma_{bars} T^* \cdot e. \tag{5.2}$$

Furthermore, R only appears on the right-hand side. For its direct solution, we must ensure the left-hand side is zero despite non-zero real displacement terms. We therefore set $W^* = 0$ and omit any virtual loading.

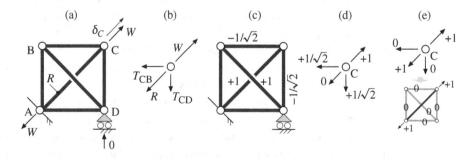

Figure 5.1 (a) Indeterminate truss where bar AC is declared to be redundant. (b) Equilibrium of joint C. (c) Bar tensions arising from self-stress when $R = 1$ and $W = 0$. (d) Equilibrium of joint C when diagonal bar tension is set equal to zero. (e) Equilibrium of joint C when only the diagonal bar tension is non-zero.

Table 5.1 Solution information pertaining to Fig. 5.1.

member	T	$e\ (\times L/AE)$	T^*	$T^* \cdot e\ (\times L/AE)$
BA, CB, CD, DA	$(W - R)/\sqrt{2}$	$(W - R)/\sqrt{2}$	$-1/\sqrt{2}$	$(R - W)/2$
AC	R	$\sqrt{2}R$	$+1$	$\sqrt{2}R$
BD	$R - W$	$\sqrt{2}(R - W)$	$+1$	$\sqrt{2}(R - W)$

However, the virtual bar tensions, T^*, must be non-zero otherwise we have a zero solution throughout; this is contrived from a state of virtual self-stress because the truss is indeterminate. We can assign any value of tension to a given bar and calculate the others from equilibrium, but it is convenient to use what exists already.

If we set $R = 1$ and $W = 0$, the tensions in Table 5.1 can be recycled into T^* and thence the product $T^* \cdot e$: see also Fig. 5.1(c). The sum, $\Sigma T^* \cdot e$, tallies the final column in the table with a fourfold first entry for four bars: $4 \times (R - W)/2 + \sqrt{2}R + \sqrt{2}(R - W)$, all times L/AE. Setting equal to the left-side zero of Eq. (5.2), we find $R = W/\sqrt{2}$.

The calculation for δ_C must use the real compatible bar extensions, e, so we appoint a virtual unit load at joint C in the same direction and use Eq. (5.2). The left-hand side reads $1 \cdot \delta_C$ with no other terms because the reaction forces at joints A and C do no work.

Our unit load equates to $W = 1$, which we can substitute into T in Table 5.1 for a corresponding set of virtual bar tensions, remembering that $R = W/\sqrt{2}$. But this creates a longer summation calculation of $T^* \cdot e$ because all bar tensions are non-zero. When we consider joint C, for example, three bars are connected to it, but we can only resolve in two directions. We can, therefore, *specify* one of the tensions before calculating the other two: and, if possible, look for the simplest solution set possible.

The selection in Fig. 5.1(d) sets one diagonal bar tension to be zero. The other virtual tensions follow from nodal equilibrium or by recycling the tensions in Table 5.1 with $R = 0$ and $W = 1$. Their values, T_1^*, are given in Table 5.2 along with their product $T_1^* \cdot e$; substituting into Eq. (5.2):

$$1 \cdot \delta_C = \left[4 \times \underbrace{\frac{W - R}{2}}_{\text{BA, CB, CD, DA}} + \underbrace{0}_{\text{AC}} - \underbrace{\sqrt{2}(R - W)}_{\text{BD}} \right] \times \frac{L}{AE}$$

$$\rightarrow \quad \delta_C = \frac{WL}{AE} \tag{5.3}$$

after replacing R, re-arranging and tidying up.

The null entry for bar AC marginally subtracts from the effort, but the simplicity of the final result hints at a more economical equilibrium set. Returning to equilibrium of joint C in Fig. 5.1(e) we can set the outer bars to be zero instead, giving a diagonal bar tension of unity.

Table 5.2 Alternative sets of virtual bar tensions for the truss in Fig. 5.1.

member	T_1^*	$T_1^* \cdot e \, (\times L/AE)$	T_2^*	$T_2^* \cdot e \, (\times L/AE)$
BA, CB, CD, DA	$1/\sqrt{2}$	$(W-R)/2$	0	0
AC	0	0	+1	$\sqrt{2}R$
BD	-1	$-\sqrt{2}(R-W)$	0	0

All the other virtual bar tensions turn out to be zero either from nodal equilibrium or by realising that the tension in AC can be transmitted directly to the grounded support at A, Fig. 5.1(e); the list of T_2^* in Table 5.2 is the simplest it can be. The Virtual Work calculation is now trivial, with $1 \cdot \delta_C = 1 \cdot \sqrt{2}RL/AE$.

This gives the same result, as we should expect (with $R = W/\sqrt{2}$), but that it does is remarkable because the complexions of the virtual sets, T_1^* and T_2^*, are so different. No matter what tensions we contrive, provided they satisfy equilibrium with the applied loading, the same result follows.

5.2   Example Two

The hanging truss in Fig. 5.2(a) has one less bar compared to before and is loaded by a pair of vertical forces, W, at its lowest joints B and C. Initially, the assembled truss fits together perfectly before loading. We wish to find the bar tensions and the vertical displacement of joint B, δ_B. We then consider what happens when bar BC is slightly longer than intended before the truss is assembled.

Nominal lengths are indicated and the geometry is straightforward: bars AC and BD are inclined at 30° to the horizontal and bars AB and CD are inclined at 60°; bar BC is horizontal. One redundancy can be verified, which we set to be the tension, R, in BC because this preserves the symmetry of calculation given the symmetrical loading.

From equilibrium of either joint B or C, we find all real bar tensions, T, in terms of W and R, which are summarised in Table 5.3. The real bar extensions, e, follow before imposing our unloaded state of virtual self-stress, in order to compute R alone – as we did previously. This is equivalent to setting $W = 0$ and $R = 1$ in Table 5.3, whence T^*. Substituting the final column of $T^* \cdot e$ into Eq. (5.2), where W^* is zero, returns:

$$0 \cdot \Delta = \left[\underbrace{2 \times \frac{\sqrt{3}W + R}{4}}_{AB,CD} + \underbrace{2 \times \frac{-3W + 3\sqrt{3}R}{4}}_{BD,AC} + \underbrace{R}_{BC} \right] \times \frac{L}{AE}$$

$$\rightarrow \quad R = \frac{2 - \sqrt{3}}{\sqrt{3}} \cdot W. \tag{5.4}$$

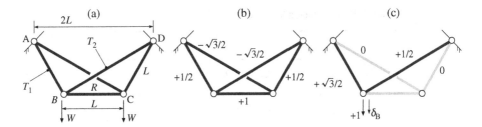

Figure 5.2 (a) Indeterminate truss with bar BC as redundant. (b) Corresponding state of self-stress. (c) Virtual equilibrium set regarding vertical displacement of joint B.

Finding δ_B mirrors earlier. It is compatible with the real bar extensions, and we impose unit force at B in the same direction, giving $1 \cdot \delta_B$ alone on the left of Eq. (5.2).

Because of indeterminacy, we use the simplest virtual bar tensions in equilibrium with the unit load. In Fig. 5.2(c), three bar tensions can be set to zero with the remaining $T^*_{BD} = 1/2$ and $T^*_{AB} = \sqrt{3}/2$. Our Virtual Work equation returns, after substituting for R:

$$1 \cdot \delta_B = \frac{\sqrt{3}}{2} \cdot e_{AB} + \frac{1}{2} \cdot e_{BD} = \left[\frac{\sqrt{3}}{2} \left(\frac{\sqrt{3}W + R}{2} \right) + \frac{1}{2} \left(\frac{\sqrt{3}W - 3R}{2} \right) \right] \frac{L}{AE}$$

$$\rightarrow \quad \delta_B = \frac{4 - \sqrt{3}}{2} \cdot \frac{WL}{AE}. \tag{5.5}$$

When bar BC is slightly longer than L, the other bars are assumed to retain their original, exact lengths. If the truss is built in sequence with bar BC connected last of all, it will not fit between joints B and C, which are an exact distance L apart.

The two halves of structure ABD and ACD are statically determinate and hence stiff, so any attempt to prize joints B and C apart before inserting bar BC is resisted elastically. If, instead, bar BC is compressed to a length L before connecting, the compressive force is transmitted to B and C and throughout the truss upon releasing.

The same response would be observed in the original, perfectly-built case if we are able to heat bar BC separately; it tries to extend in-situ but is constrained, thereby generating bar tensions. The following method of solution, therefore, applies to heating/cooling problems as well as those involving nominally *mis-fitting* bars just too short or too long.

There is clearly a state of self-stress, and we declare the bar tension in BC to be the redundancy, which allows us to recycle the previous *real* bar tensions from Table 5.3 with $W = 0$. All bar tensions depend on R, so do the real bar extensions, but only one value of R leads to a compatible geometry of deformation.

There is clearly a state of self-stress, and we declare the bar tension in BC to be the redundancy, which allows us to recycle the previous *real* bar tensions from Table 5.3 with $W = 0$. All bar tensions depend on R, so do the real bar extensions, but only one value of R leads to a compatible geometry of deformation.

Table 5.3 Solution information pertaining to Fig. 5.2.

member	length	T	$e\ (\times L/AE)$
AB, CD	L	$(\sqrt{3}W + R)/2$	$(\sqrt{3}W + R)/2$
BD, AC	$\sqrt{3}L$	$(W - \sqrt{3}R)/2$	$(\sqrt{3}W - 3R)/2$
BC	L	R	R

member	T^*	$T^* \cdot e\ (\times L/AE)$
AB, CD	$+1/2$	$(\sqrt{3}W + R)/4$
BD, AC	$-\sqrt{3}/2$	$(-3W + 3\sqrt{3}R)/4$
BC	$+1$	R

Table 5.4 Solution information for Fig. 5.2 when bar BC is too long by an amount λ.

member	T	e	T^*	$T^* \cdot e$
AB, CD	$R/2$	$RL/2AE$	$+1/2$	$RL/4AE$
BD, AC	$-\sqrt{3}R/2$	$-3RL/2AE$	$-\sqrt{3}/2$	$3\sqrt{3}RL/4AE$
BC	R	$RL/AE + \lambda$	$+1$	$RL/AE + \lambda$

The extensions are given in Table 5.4 where BC has an extra term due to its initial over-length, denoted as λ. This is treated as an extensional term because if we imagine disconnecting the bars, there can be no bar forces and R is zero but BC remains extended beyond L.

To find R, we impose a unit redundancy on BC, *i.e.* $R = R^* = 1$ and no other loads, returning zero to the left-hand side of Eq. (5.2). Substituting R gives the other virtual tensions and their product, $T^* \cdot e$, in the final column of Table 5.4. Summing in accordance with the right-hand side of Eq. (5.2), we find:

$$0 \cdot \Delta = \underbrace{2 \times \frac{RL}{4AE}}_{\text{AB, CD}} + \underbrace{2 \times \frac{3\sqrt{3}RL}{4AE}}_{\text{BD, AC}} + \underbrace{\frac{RL}{AE} + \lambda}_{\text{BC}}$$

$$\rightarrow \quad R = -\frac{2}{3(1 + \sqrt{3})} \cdot \frac{EA\lambda}{L}. \tag{5.6}$$

The compression of bar BC, as if 'shoe-horned' into place during assemblage, is confirmed by the minus sign. Writing λ in terms of RL/AE, we find a pre-factor equal to $-3(1 + \sqrt{3})/2 \approx -4$. Looking at the bar extension for BC in Table 5.4, the λ component is thus four times larger (and negative) compared to that due to R alone – the *elastic* component due to force.

Bar BC is, therefore, not *entirely* compressed to a length L, which could only happen if it was fitted last of all to a perfectly made, rigid truss elsewhere. The original length, $L + \lambda$ decreases to around $L + 3\lambda/4$, which is permitted by elastic adjustment throughout the truss.

5.3   Example Three

The simple triangulated truss of equal length bars, L, in Fig. 5.3(a) is supported by an unusual but feasible set of roller supports: at A, the support is vertical, at B, horizonal, and at C inclined at $30°$ to the horizontal.

At each support, there is, therefore, a single normal reaction, and Maxwell's Rule confirms determinacy: $b + r - D \cdot j = 5 + 3 - 2 \times 4 = 0$. No loads are applied, and bar AB extends by a small amount e compared to L: we wish to find the displacement of roller C, δ_C.

On first sight, the imposition of an isolated bar extension can be unsettling. We expect extensions from some kind of force loading, pervading the truss; however, we can imagine heating only bar AB (if made of metal) with no loads applied, which causes it to extend during expansion – as per the previous lack of fit. But unlike before, there is no state of self-stress because of statical determinacy, which simplifies our solution.

The single real extension also reduces the non-zero detail on the right-hand side of Eq. (5.1), and whatever virtual set of tensions we work with, only T_{AB}^* will contribute to the calculation. We now calculate this tension from two loadings: the standard approach and a variant of this to reduce the calculation involved.

In order to make δ_C explicit, it is usual to apply a virtual unit force in this direction, Fig. 5.3(b). As the free body of joint C attests in Fig. 5.3(c), we have three unknowns, the reaction R_C^*, and the bar tensions T_{CD}^* and T_{CB}^*. For two equations of force equilibrium, these are insoluble.

We return to Fig. 5.3(b) and observe moment equilibrium about joint B for the entire truss. The reaction forces at joints A and B do not contribute but R_C^* does alongside the unit force at C, giving $R_C^* = 1/\sqrt{3}$. Going back to Fig. 5.3(c) and resolving in two directions, we eventually find $T_{CB}^* = 2/\sqrt{3}$ and $T_{CD}^* = 0$. The zero result is not obvious any other way, but it tells us to expect zero bar tensions all around joint D.

We can now resolve horizontally for the entire structure to find R_A^*, which is equal and opposite to T_{AB}^* given that T_{AD}^* is zero; or we can find T_{AB}^* by moving from equilibrium of joint C to joint B. The final bar tensions and reactions are displayed in Fig. 5.3(d) for completeness, and Eq. (5.2) now returns: $1 \cdot \delta_C = T_{AB}^* \cdot e + T_{CB}^* \cdot 0$ i.e. $\delta_C = e/\sqrt{3}$.

We can simplify the analysis by applying a unit force at joint C *parallel* to bar CB, Fig. 5.3(e). The inclination of force is different to the displacement, but importantly it exerts no moment this time about joint B, and R_C^* is immediately zero from global moment equilibrium. Force equilibrium at joint C sets T_{CB}^* equal to the unit force and T_{CD}^* to zero, and thus $T_{AD}^* = 0$ and $T_{DB}^* = 0$ around joint D. Given the latter, $T_{AB}^* = 1/2$, from equilibrium at B.

The right-hand side of Eq. (5.2) is thus $(1/2) \cdot e$ and the left must engage the dot product because the unit force and δ_C are not parallel, giving us $(1 \times \cos 30°) \cdot \delta_C$. Equating both terms and re-arranging, we arrive at the same result $\delta_C = e/\sqrt{3}$.

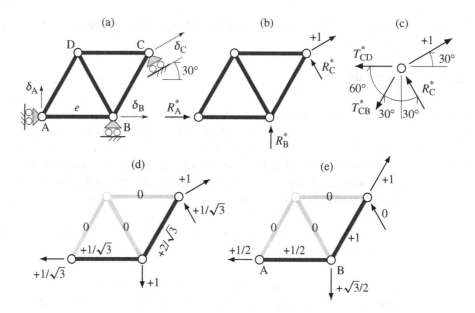

Figure 5.3 (a) Determinate truss with three roller supports: bar AB extends by e, all other bars remain at length L. (b) Virtual forces applied to the truss. (c) Corresponding equilibrium of joint C. (d) Overall bar forces and reactions. (e) Alternative virtual equilibrium scheme with unit force applied parallel to bar BC.

Even though the level of working is similar to the first case, there are more zero terms, making the calculation simpler and less prone to errors. We modified the load direction from that of the real displacement because we anticipated an equilibrium benefit rather than declaring the usual direction and calculating the bar tensions passively.

5.4   Final Remarks

A good choice of redundant bars can reduce the level of calculation availed by Virtual Work. The middle lower bar in Fig. 5.2 connects to symmetrical *sub-structures* on either side, which have the same statical properties from the overall loading – and only one side needs to be evaluated. Declaring the diagonal bar in Fig. 5.1 to be redundant enabled the general set of equilibrium values to be recycled at various calculation stages.

However, the redundant bar in Fig. 5.2 offers a more subtle point. A *particular* equilibrium solution arises in Table 5.4 from R taking *any* numerical value for the loads applied. If zero is set, the sub-structures are effectively disconnected from each other, and whatever the applied loading, their bar tensions are automatically easier to calculate. Thus, we seek out redundant members that connect simpler but isolated sub-structures.

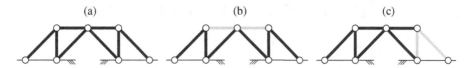

(a)   (b)   (c)

Figure 5.4 (a) Symmetrical truss. (b) Two redundancies declared as the top bars, leaving three separate, statically determinate sub-structures around them. (c) Inadmissible choice of redundant bars, leaving a local mechanism.

Symmetry, as usual, complements this assessment. For example, the eight-bar truss in Fig. 5.4(a) has two redundancies, $b + r - D \cdot j = 8 + 4 \times 2 - 2 \times 7 = 2 = s$. The pair of horizontal bars are prime redundancy candidates, connected to three separate sub-structures – two being the same whatever the applied loading, Fig. 5.4(b).

But we cannot 'extract' any two redundant bars in this thought process; whatever partial structure is left behind should still be determinate without local mechanisms. The redundant declaration in Fig. 5.4(c) leaves the top bar 'unsupported', which is inadmissible.

Finally, the flip side of Virtual Work is equally important; of finding displacements or bar extensions. Indeterminate structures allow us to select the simplest virtual equilibrium set, ideally with as many zero bar tensions as possible, making $\Sigma T^* \cdot e$ easier to compute. This is the strength of the Virtual Work method, and can be verified by choosing more than one virtual equilibrium set for the same problem: the same real displacement or bar extension will emerge.

6   Why Does the Lower Bound Theorem Work?

Rather, why should *any* equilibrium solution which does not violate permanent yielding of the material afford safe working of a structure? The answer is, of course, imbued by the mathematical formalism of the Lower Bound Theorem, which we now discuss informally.

Application of the Lower Bound Theorem relies upon ductility, where material yielding permits enough plastic straining for a pronounced structural manifestation. Metals are fortuitously, and mostly, ductile, but other Engineering materials, *e.g.* concrete, have to 'contrive' ductility in special cases, namely when there is insufficient metal reinforcement, which yields before catastrophic brittle failure of the concrete. Crucially, whatever the mechanism of yielding for our material, its ductility proceeds at *constant* stress.

A (metal) bar under tension will yield everywhere uniformly, giving a simple connection to structural ductility. Yielding in bending, however, is more complex, first occurring at the outermost fibres of the cross-section. Sustained bending causes plastic deformation to encroach from the top and bottom of the section towards the neutral axis, which strictly becomes the equal areas axis in the limit (see Chapter 15).

Determining the precise extent of plastic deformation must consider all three structural imperatives, but eventually full plasticity is achieved from being ductile. In practice though, excessive deformation tends to *localise* rather than to pervade. In tensile bars, we see familiar *necking* at a given cross-section, and in bending, most of the beam off-loads away from concentrated deformation into a so-called *plastic hinge*.

But therein lies the paradox: why is ductility, which implies significant yield strains, a prerequisite for applying the Lower Bound Theorem, in order to return a safe loading limit where the structure has not collapsed, *i.e.* not yielded critically (or entirely)?

First, we note that a safe load does not necessarily invoke a purely elastic response; part or parts of the structure can have yielded already without substantive deformation and no local collapse. As the loading rises, elastic stresses elsewhere build up even though the yielded regions have saturated in terms of stress levels – but not strain levels from being ductile. This process of *redistribution*, as it is known, maintains the capacity for the load increasing up to its *ultimate* value, just before collapse occurs.

A statically determinate structure is governed, however, by a single equilibrium solution. Its internal stresses are known exactly when material yielding begins, which defines a single limit of safe loading without any redistribution. Geometrical compatibility is automatically and separately satisfied.

Redistribution is thus a feature of how indeterminate structures respond, and is captured analytically from knowing that infinitely many equilibrium combinations are viable (*c.f.* Chapter 5). Assessing geometrical compatibility adds nothing to understanding the limit of safe loading; of course, it tells us about the level of straining which ductility ensures is a non-critical feature as loading progresses.

Thus, we do not have to assess compatibility, returning an economy of working. In return for whatever viable equilibrium solution we propose, the ultimate load must be a safe value – as in our original question – and is underwritten by not violating the yielding condition.

At one extreme, where all stresses are elastic, this is automatic. But often we designate parts of the structure proximal to yielding for a more accurate ultimate load prediction. For simple enough structures, the number and location of these parts will be obvious – from the loading response profile. Alternatively, we may appeal to the Upper Bound Theorem for a sense of the collapsed state, which also originates from these locations.

Just enough locations are on the brink of actual yielding, and our proposed equilibrium solution, therefore, approaches their specification (if it is not the correct one). There will potentially be some redistribution available from our proposition and thus some residual loading capacity, *i.e.* our ultimate prediction will always be less than (or equal to) the true value, and, therefore, safe.

Even though many studies introduce Lower Bound in the content of beams and frames, we now illustrate the above remarks using a singly-redundant truss structure for its relative economy of analysis; in Chapter 16, we deal with beams and frames.

Because we are potentially dealing with slender bars in compression, they *cannot* be susceptible to buckling, for this undermines true ductility and is why Lower Bound design of trusses is rarely expounded. A separate elastic analysis is required for the exact bar forces, in order to deal with buckling.

However, our ultimate aim is an exact response including plastic effects to compare against, and our simple example contrives tensile bar forces throughout, thereby precluding buckling. We then compare this response to our proposed Lower Bound solutions, paying particular attention to the exact deformation in order to extol the role of ductility.

6.1   Exact Behaviour

In Fig. 6.1(a.i), bars are either L or $\sqrt{2}L$ long, have the same cross-sectional area A and Young's Modulus E. The symmetry of layout and the force, F, applied horizontally ensure that the outer bar tensions are the same and equal to T_1: the middle bar tension is T_2. Equilibrium of the loaded pin-joint in Fig. 6.1(b) gives us a single statement for two unknown bar tensions: $F = \sqrt{2}T_1 + T_2$.

The same joint displaces by amount d horizontally. During the initial elastic response, the bar extensions are defined by $e_1 = T_1(\sqrt{2}L)/AE$ and $e_2 = T_2L/AE$. The extension e_2 is equal to d by definition: the other bars also rotate as well as extend

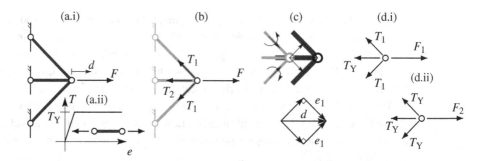

Figure 6.1 (a.i) Indeterminate symmetrical truss and horizontal loading; (a.ii) constitutive response of a given bar up to yield in tension. (b) Equilibrium of the loaded joint. (c) Kinematics of loaded joint and displacement diagram. (d.i) First yield phase; (d.ii) Total yield.

with small displacement components respectively normal to and along the bar. The corresponding vector diagram in Fig. 6.1(c) shows that $e_1 = d/\sqrt{2}$. The ratio e_2/e_1 is thus $\sqrt{2}$; therefore, $T_2/T_1 = 2$.

The same ratio can be found as quickly by using the Virtual Work method of Chapter 5. Because of the indeterminate nature, we can set either virtual tension to be zero provided we satisfy equilibrium, in order to solve directly for the other. Setting $F^* = 1$ and T_2^* to be zero, T_1^* equals $1/\sqrt{2}$, and with e_1, e_2 and d as our real quantities, Virtual Work states that:

$$\text{Eq. (5.1)} \rightarrow 1 \cdot d = 2T_1^* \cdot e_1 + T_2^* \cdot e_2 = 2 \times \frac{1}{\sqrt{2}} \frac{\sqrt{2}T_1 L}{AE} + 0$$

$$\rightarrow \quad T_1 = \frac{AE}{2L} \cdot d. \tag{6.1}$$

Setting T_1^* to be zero, T_2^* equals unity for $F^* = 1$, and Virtual Work again returns $T_2 = (AE/L) \cdot d$.

The elastic response considers how F and d are related. Substituting both tensions into the original equilibrium equation and re-arranging non-dimensionally:

$$F = \sqrt{2}T_1 + T_2 \quad \rightarrow \quad F = \frac{\sqrt{2} + 1}{\sqrt{2}} \cdot \frac{AE}{L} \cdot d \tag{6.2}$$

and the displacement builds up linearly as the load increases, until yielding is reached.

Because T_2 is larger, the middle bar yields first. The limiting tension is given by assuming that the cross-section has yielded uniformly, as noted earlier, with a value denoted by T_Y equal to A multiplied by the yield stress of the material, σ_Y. Even though we assume a perfectly plastic response in this phase, which maintains σ_Y irrespective of the level of plastic straining, Fig. 6.1(a.ii), we expect localised necking if the strains become excessive in practice.

Equilibrium of the loaded joint in Fig. 6.1(d.i) now gives $F = \sqrt{2}T_1 + T_Y$, with further loading increments only from T_1 increasing elastically: the original elastic ratio for $T_2/T_1 = 2$ is no longer valid. Eventually, the outer bars also yield where

F equals $\sqrt{2}T_Y + T_Y$ from Fig. 6.1(d.ii). This is the ultimate load with no further increase possible. The loaded joint will, of course, continue to displace if the load is maintained.

There are three phases of behaviour: elastic (E), partially plastic, *i.e. elasto-plastic* (EP) and fully plastic (P). Defining dimensionless terms $f = F/T_Y$ and $t_{1,2} = T_{1,2}/T_Y$, we can write our found equilibrium statements for each phase more compactly:

$$(E): \; f = \left[2 + \sqrt{2}\right]t_1, \;\; t_2/t_1 = 2 \;\; (a),$$

$$(EP): \; f = \sqrt{2}t_1 + 1, \;\; t_2 = 1 \;\; (b),$$

$$(P): \; f = 1 + \sqrt{2}, \;\; t_1 = t_2 = 1 \;\; (c). \tag{6.3}$$

These phases are captured in the tension-force plot in Fig. 6.2(a) (t_1 is the black line, t_2 is grey) starting off with the elastic (E) phase.

This phase ends when expressions (E) and (EP) are equal to give first yielding at $f = 1 + 1/\sqrt{2}$ (=1.71), $t_1 = 0.5$ and $t_2 = 1$, which are plotted as solid circles. For larger values of f, t_2 remains at unity, and t_1 from (EP) increases linearly up to the same at a final value of $f = 1 + \sqrt{2}$ (=2.41). At this point, all bars have yielded and we have the final ringed point for the plastic (P) phase.

We also plot the corresponding load-displacement (F, d) profile in Fig. 6.2(b). Equation (6.2) already defines the elastic response: if we divide both sides by the strain at first yield, $\epsilon_Y = \sigma_Y/E$, the left-side denominator becomes $F/A\sigma_Y$, which is f. On the right-hand side, the new ratio $d/L\epsilon_Y$ is defined as δ, and is the ratio of the equivalent strain from the joint displacement, (d/L), to yield strain.

The elastic strain in the middle bar reaches ϵ_Y when $\delta = 1$ (and $f = 1 + 1/\sqrt{2}$), which marks the end of the elastic phase. During the elasto-plastic phase, we note that T_1 is as per Eq. (6.1).

The Virtual Work calculation, which is elastic by nature, still applies in this phase even though T_2 has yielded because our virtual T_2^* has been set to zero. If we convert f and t back to the actual ratios, substitute for d from T_1, and then reinstate f and t, we find:

$$\frac{F}{T_Y} = 1 + \frac{\sqrt{2}T_1}{T_Y} = 1 + \frac{1}{\sqrt{2}\epsilon_Y} \cdot \frac{d}{L} \;\; \rightarrow \;\; f = 1 + \frac{\delta}{\sqrt{2}} \tag{6.4}$$

after noting that $AE/T_Y = A\,E/A\sigma_Y = 1/\epsilon_Y$.

This second linear curve has a smaller gradient, hence stiffness, because the middle bar has reached its tensile limit: the unity intercept enables the non-zero intersection of these first two phases. The elasto-plastic phase ends when $f = 1 + \sqrt{2}$ equals $1 + \delta/\sqrt{2}$ from the plastic phase: accordingly, $\delta = 2$, which is twice the yield strain. There is also no stiffness in this final phase, and displacements increase under constant f.

During this phase, necking is unlikely given the low level of plastic straining, but if the ultimate load is maintained or we try to apply a larger load, the loaded joint will suffer runaway displacements with eventual necking of the bars.

In practice, we ensure that the actual load applied is smaller than the dimensioned ultimate load, which is $(1 + \sqrt{2})A\sigma_Y$. Most likely, σ_Y will be fixed by a single choice of material but A will take discrete values from a 'catalogue' of available cross-sections. Ever-increasing ultimate loads can be specified using larger areas, resulting in a heavier truss: a sensible final design may have to be a compromise between mass and capacity.

6.2   Lower Bound Trials

Even for this simplest of indeterminate trusses, our working has not been trivial to account for exact displacements and true equilibrium tensions. The ultimate loading, however, turns out to be a simple equilibrium calculation when both bar tensions are set to be T_Y, and the benefit of considering equilibrium solutions by themselves is perhaps obvious.

These final bar tensions should, however, be on the cusp of reaching T_Y so that the yielding condition is not strictly violated. Any other limiting equilibrium solution will obviously have either of T_1 or T_2 just equal to T_Y but not both, giving a lower ultimate load compared to the optimal value. For example, if we set $T_1 = T_Y$ and $T_2 = 0$, the ultimate load is $F = \sqrt{2}T_1 + T_2 = \sqrt{2}T_Y$ or $f = \sqrt{2}$.

This result is plotted as the first dashed vertical line in Fig. 6.2(a). The corresponding true tensions from the intersection points are non-zero and can be read off or calculated from Eqs. 6.3 to give $t_1 = 1/(1 + \sqrt{2}) = 0.414$ and $t_2 = 0.828$. Being less than unity, both tensions are elastic and our structure is safe but operating well below its ultimate capacity.

Imagine, however, that we are ignorant of the true ultimate load but wish to improve our conservative trial solution with a higher, elastic value for T_2 whilst retaining T_1 at T_Y. In particular, set T_2 conveniently to $(2-\sqrt{2})T_Y (= 0.585T_Y)$, which improves our load estimate, now $F = \sqrt{2}T_1 + T_1 = 2T_Y (f = 2)$.

When plotted as another dashed line in Fig. 6.2(a), $(f = 2)$, the true tensions, in fact, behave oppositely: $t_1 = 1/\sqrt{2} = 0.707$, which is elastic, and $t_2 = 1$, at yielding.

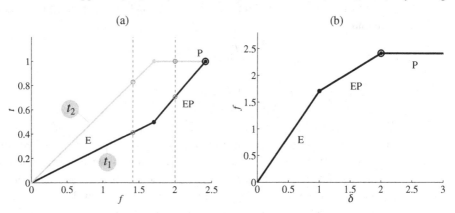

Figure 6.2 (a) Tension-load response of the truss from Fig. 6.1. Solid lines are exact bar tensions; solid circles mark the end of the elastic (E) phase and the start of the elasto-plastic (EP) one. The intermediate ringed values via the dashed lines are Lower Bound solutions; the final ringed value is the plastic (P) phase. (b) Corresponding load-displacement response.

Furthermore, the intersection with the t_2 line is past the point at which it first yields (recall when $f = 1.71$). From Fig. 6.2(b), $\delta = \sqrt{2}$ when $f = 2$, giving a middle bar strain of $\sqrt{2}\epsilon_Y$.

Clearly, the middle bar has yielded appreciably by the time $f = 2$ despite assuming an elastic tension. Such disparity is *protected* by material ductility; if instead the bars were made from a brittle material which failed, say, at a strain equal to ϵ_Y, the middle bar would break at a load of $f = 1.71$, some way below the safe load prediction. Equilibrium of the loaded joint at this point discards T_2 to give $f = \sqrt{2}t_1$, which returns $t_1 = 1.71/\sqrt{2}$, greater than unity, *i.e.* the outer bar would instantaneously fail also.

6.3 Final Remarks

Ductility provides a literal safety margin for our 'incorrect' trial solutions compared to true behaviour, and coupled to the Lower Bound Theorem returning a smaller, safe ultimate load every time, the freedom to select equilibrium solutions is key to its efficacy.

If we are designing the truss for a fixed load in practice, a lower ultimate prediction means using bars with larger cross-sectional areas than is necessary. The truss, we say, will be *over-designed* in terms of being able to accommodate a higher fixed load (which hopefully is not applied in practice).

This final statement speaks to the remarkable prescience of the Lower Bound Theorem in general as a design tool. Where else, in Engineering, does an incorrect solution result in a margin of safety and increasingly so for a greater departure from the exact solution? The detriment is, of course, a more massive structure.

We will invariably over-design more complex structures, which have many more equilibrium combinations. Rather than trying to seek one of the better ones, we should aim for simplicity if pressed for time where a 'back-of-the-envelope' calculation is not anachronistic.

Again, we note that only the final equilibrium state matters; compatibility does not have to be assessed because of ductility, which is simply contrived. In practice, yielding leads to higher stresses after initial yielding, which would result in the junctions between phases in Fig. 6.2 becoming more rounded, for example.

We also have a perfect structure in which bars neatly connect together without mis-fitting. There will inevitably be small misalignments with some bars having to be stretched or compressed in order to complete building the truss, giving non-zero tensions for zero loading, as we saw in Chapter 5.

Any state of self-stress increases the proximity *to* yielding but does affect the final capacity of individual members. In this regard, deliberate pre-stressing and unforeseen movements of supports do not subtract from the ultimate load.

Part III

Beams and Frames: Character

7 Drawing Well: Bending Moment (and Shear Force) Diagrams

Often the scourge of undergraduate exercises, drawing bending moment profiles for slender structures is an essential skill. Their variation tells us about how a beam or column is loaded transversely as well as indicating salient values at pertinent locations where damage may be incurred from extreme loading.

In the case of, say, a steel beam, it will submit to permanent yielding of the material at these points, which concentrates into plastic hinges; collapse of the entire structure may follow depending on how the structure adjusts to more loading. Equally important are shear force diagrams – especially for concrete structures, where failure in shear often dominates that of bending.

The usual way to teach construction of bending moment (and shear force) diagrams is through piece-wise equilibrium. We select a given free body, resolve forces and take moments to establish its variation before doing the same with a different free body. We then assemble the complete profile reliably but inefficiently. We shall adopt a more holistic approach.

For small variations in bending moment, M, and shear force, S, across an element of straight beam of length δx, Fig. 7.1(a), equilibrium gives their well-known relationships to some externally applied, transverse loading intensity, w, as:

$$\frac{\mathrm{d}S}{\mathrm{d}x} = w \quad \text{(a)}, \quad \frac{\mathrm{d}M}{\mathrm{d}x} = S \quad \text{(b)} \quad \rightarrow \quad \frac{\mathrm{d}^2 M}{\mathrm{d}x^2} = w \quad \text{(c)}. \tag{7.1}$$

These statements convey much information.

First, the directions of M and S obey an explicit sign convention, which we must *always* declare *a priori*. Here, M is positive if the beam is locally *hogging* and curving upwards, and shear forces point downwards on the left side and *vice versa* right-side. The latter is also linked to the direction of increasing x, for if x is reversed in measurement along the beam, so do the positive shear force directions, in order to preserve the statements above. Indeed, our choice in absolute directions for all leads to an absence of minus signs when w acts downwards in the sense of gravity. We can, of course, choose a different scheme, such as one with positive sagging bending moments, which will introduce minus signs.

We can find the shear force profile by differentiating that of bending moment, if it is constructed first. If not, we may integrate shear to find bending moments, which is tantamount to finding the area underneath the shear force profile. This is a slightly more challenging mental performance notwithstanding any, as yet unknown, constants of integration and how to interpret them; it is marginally easier to deal with bending moments first.

When the self-weight is uniform, w is a constant and M is quadratic, with two constants of integration usually equal to known values of bending moments at supports, joints or junctions. For any other variation in w, we integrate the same, but we note that for w specified as a polynomial function, M is a polynomial of two orders higher; recall our previous experiences with the shape of loaded cables, Chapter 3.

However, when w is zero, the complementary function of Eq. (7.1)(c) is linear, which must correspond to the bending moment profile between point loadings or reactions, both for applied forces *and* moments. At the position of a given force, we ascertain the bending moment from overall equilibrium but either side, the slope is different – as in a point-loaded cable. This is because the shear force, *viz.* the gradient, changes value, most evidently from equilibrium of a small element in Fig. 7.1(b).

We only show forces and resolve vertically, where we find the right-side shear force, S_2, equal to S_1 (left) plus the applied force, F. As the width tends to zero, there is a discontinuous step up in shear force, equal in magnitude to F. The direction of the step should always be clarified this way each time for complete assurance.

For an applied couple or torque, C, a small element shows in Fig. 7.1(c) a similar performance, of a jump in bending moment, $M_2 - M_1$, equal in value, which we draw as a step change in the limit of zero width. Interestingly, there is no change in shear force for this case – as there is no change in bending moment about a point force.

The bending moment is always zero at a pin-joint because the beam portions on either side can rotate freely relative to each other. Relative translations are not possible otherwise the pin tears apart: assuming that it behaves rigidly in these senses, it can 'transmit', *i.e.*, carry both shear forces and axial tensions.

A general connection from one beam to the next may be obliged to transmit all three quantities; transmitting a moment by itself is, however, a less common requirement but it is practicable, as we shall see.

As a last point, there are step changes in shear force at simple supports, which are especially relevant to convey at the ends of the beam, where often they are not; we can think of continuing the beam a little way beyond this final support (as it would be in practice), to emphasise the step up or down to the datum line, as in Fig. 7.1(d).

We now consider a number of worked examples to reinforce these ideas. Note that Eq. (7.1) is valid also when the beam is gently curved where x is an intrinsic coordinate for its particular shape. If substantially curved like an arch, axial forces are an inevitable loading response alongside possible coupling to bending moments (unlike, say, the 'portal frame' from later whose straight members preclude such interaction); this is presently beyond our scope.

7.1   Moments and Cables

When we compare Eqs. (3.3) and (7.1), the equilibrium displacements of a cable, y, and distribution of bending moments, M, are equivalent, save for a factor of H

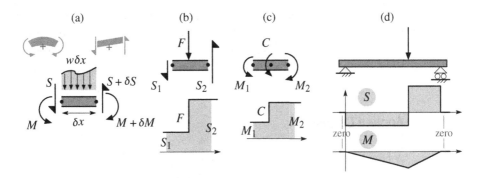

Figure 7.1 (a) General variation of bending moment M and shear force S across an element of straight beam under transverse loading intensity w. (b) Applied point force F causes a step change in shear force. (c) Similarly, a jump in bending moment is produced by an applied couple or torque. (d) Profiles for a simply-supported beam: note how they overrun the ends, for completeness.

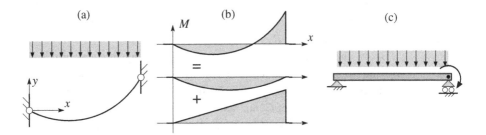

Figure 7.2 (a) Uniformly loaded, asymmetrical cable with a quadratic equilibrium profile. (b) Similar bending moment variation, separated into linear and uniform profiles. (c) Corresponding loading on an equivalent simply-supported beam.

(the constant horizontal reaction force). We have an *analogy* for describing either system in terms of the other, but for us, we want to see if the former can help us sketch the latter more effectively.

We re-deploy the asymmetrical cable under uniform loading w from Fig. 3.1 in Fig. 7.2(a). We argued then for a quadratic profile, which becomes our proposed bending moment diagram in Fig. 7.2(b); we now ask what loading does this represent on the beam?

First, we must ask what is the beam? Both sets of governing equations pertain to a (vertical) loading intensity normal to a linear coordinate, x; this is the beam centre-line, and the intentional axis in Fig. 7.2(b). We set the left-hand side moment to be zero because that is where the cable origin lies; if the origin moves up or down, the bending moment profile shifts accordingly with a non-zero offset.

The right-side moment must step down to zero moving past where the right cable pin is, and conforms to an external moment applied to same end of the beam. We can uncouple the quadratic profile into linear and symmetrical components, as shown, for an end moment and a uniform loading intensity, see Fig. 7.2(c).

The former is positive and causes the beam to hog, so the end moment must be clockwise on this right side; the latter produces sagging and acts downwards. Across both ends, the bending moment gradient changes discontinuously from zero, giving rise to applied vertical forces, which come from simple supports, for example.

7.2   Applied Couple

A couple, C, is applied externally to the simply-supported beam in Fig. 7.3(a). It is wrought in practice by its namesake, a pair of equal and opposite planar forces acting on small rigid levers attached to the beam, as shown. The exact position of the couple is not needed for a general profile.

As much *global* information as possible is established first. The couple tries to twist the beam anti-clockwise in the plane but is restrained by the vertical support reactions; on the left-side, an upwards force is applied to the beam, and right-side, it is downwards. The gentle horizontal S-shape displacements are added in Fig. 7.3(a) as a useful *aide-memoire*. There are no other forces acting on the beam, giving equal and opposite reactions, R.

Below the beam, the general features of bending moment profile are surmised. Moving from left to right, we have zero, then linear for the clear span up to the couple, and a discontinuous step, followed by linear, then ending with zero again.

Given the direction of C, our bending moment sign convention informs an upwards jump of C left-to-right, see Fig. 7.3(b). We establish absolute values at the top and bottom shortly, but we recall that the beam is bending downwards left of the couple, and *vice versa*, signifying sagging (negative) and hogging (positive) moments respectively on either side. The step, therefore, straddles the bending datum, as shown, anywhere between $\pm C$ vertically.

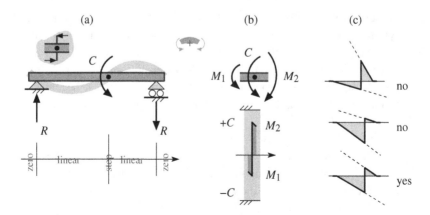

Figure 7.3 (a) Couple applied to a simply-supported beam and the proposed bending moment variation. (b) Step change in bending moment across the couple position. (c) Nature of the bending moment gradients either side of the couple, being correctly the same.

Connecting the linear parts on either side to zero at both ends, we arrive at an informal bending moment profile without formal calculation. Its shape depends on the absolute position of the step vertically, Fig. 7.3(c), so we think about gradient and shear forces.

When the step position is too high, the left-side gradient is shallower than the right *etc.*, giving a difference in shear forces across C, either way. However, no external force is applied here with no change in shear force: the gradients must therefore be equal. We can now locate the step easily to achieve this.

Figure 7.4(a.i) indicates the general bending moment profile with zero at the ends clearly highlighted. The shear force diagram is drawn below, Fig. 7.4(a.ii), and confirms the reaction force directions from equilibrium of small elements at both ends of the beam; we have drawn the shear forces in their true directions, for example, being negative and hence downwards on the left-side of the first element, with no shear force on the beam end, thus giving R upwards.

Can we also determine the original bending moment profile from some loading of some cable? Remembering that cable displacement and moment are analogous, there is no difference in height between the cable ends, which have the same horizontal span.

The cable must, however, be disjointed at the position of C on the beam; we can think of cutting the cable at this position and moving the new ends apart by an amount C vertically, stretching the cable in the process. The absolute positions of these ends are, however, not clear.

We can also apply a couple C *to the cable* by replacing part of it in the vicinity of C with a small length of rigid beam projecting symmetrically on either side, Fig. 7.4(b). It is now twisted through $90°$ by applying C in the same direction as before until it becomes vertical, giving the same disjointed cable profile, but only because the loaded beam is in equilibrium.

A free body in Fig. 7.4(c) shows that the vertical and horizontal components of tension must be equal, *i.e.* when the tensions applied to the top and bottom of the beam are parallel. Relative to the ends of the cable, this requirement sets the vertical position of the beam, and it tells us that the bending moment gradients on either side of the step must also be equal.

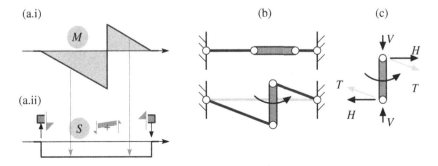

Figure 7.4 (a.i) Formal bending moment profile of the loaded beam from Fig. 7.3(a); (a.ii) corresponding shear force profile. (b) Analogous displaced cable structure after twisting a rigid link. (c) Equilibrium of forces on the link.

7.3   Two Spans

Figure 7.5(a) shows a continuous beam of two equal spans, each of length L, with a central load W applied to the first span. The beam is statically indeterminate, but we stipulate the central reaction to be a 'known' quantity λW, which expresses all other statical quantities. Only one particular reaction value, however, will yield a geometrically compatible set of elastic deformations – if we were interested; but we deal only with equilibrium solutions presently, for which λ can take any value.

An exaggerated displaced shape shows the first span dipping downwards. Most of the second span curves upwards and would lift off, which suggests a downwards reaction, R_B, at the right end. Our bending moment expectations are indicated below it: zero, through linear thrice between point loadings and reaction forces, and back to zero.

Because we are not concerned about compatibility, we can neglect the requirement of zero displacement over the middle support and treat the loading as two simply-supported cases. The bending moment profiles are now trivial, with Fig. 7.5(b.i) showing them in opposite senses due to the directions of W and λW. The peak values

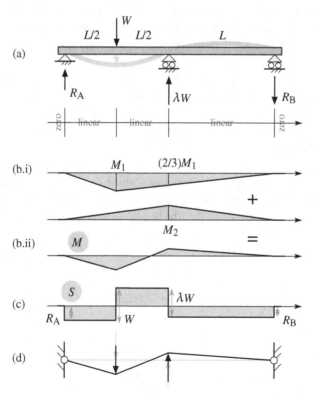

Figure 7.5 (a) Two-span beam with externally applied forces and proposed bending moment variation. (b.i) Separated moment profiles from W and λW alone without internal roller support; (b.ii) then superposed. (c) Corresponding shear force profile. (d) Analogous displaced cable.

are, respectively, $M_1 = (3W/4) \cdot (L/2) = 3WL/8$ and $M_2 = (\lambda W/2) \cdot L = \lambda WL/2$ using a free-body diagram from the left-side support to each point force.

These are superposed in Fig. 7.5(b.ii), which shows a positive bending moment in the middle: by comparing the separate salient values above, this occurs when $M_2 > (2/3)M_1$, where two-thirds arises from using similar triangles in the right-side of the M_1 profile. Consequently, $\lambda > 1/2$, which also ensures positive R_B downwards. If λ is less than one-half, the bending moment everywhere is negative and R_B is reversed.

Its gradient yields the shear force diagram in Fig. 7.5(c), which naturally expresses constant values between steps equal in value to the reaction force, R_A, and the rest of the forces. Their directions are also consistent with those in (a).

The displaced cable analogy is straightforward, for the beam is straight and horizontal, as per the initial cable: Fig. 7.5(d) clearly mimics the bending moment profile.

7.4   Transmitting Moments

The cantilever in Fig. 7.6(a) is *propped* by an internal support located two-thirds along towards the tip. The span L inside is divided into halves by a pin-joint, and a vertical force F is applied to the tip.

Once again, we exaggerate deflections to garner a sense of the bending behaviour. The second portion of the beam bends over the support, and the end at the pin displaces upwards, which drags the beam inside upwards. There is a relative rotation across the pin for no moment can be transferred across it.

The bending moment form is first given below with no surprises. The 'step' indicated at the left-side built-in support is, perhaps, a misnomer because the moment there continues well into the wall and does not step down to, or up from, zero.

If the built-in end were, in fact, free of the wall, we need to apply an external moment as well as a vertical reaction, to prevent any movement or rotation in keeping with the constraint applied by the wall. For that reason, we show the step in bending moment akin to an external moment.

Otherwise, the profile is straightforward to construct. The left-side step is negative because it counteracts any tendency for anti-clockwise rotation of the inside beam. The moment then rises linearly to zero at the pin, and maintains the same gradient beyond it, becoming equally positive as far as the internal support. After this, it changes direction and drops linearly to zero. We only need one actual value to quantify the entire scope: from a free body just beyond the internal support, the bending moment is $F \cdot (L/2)$.

The shear force profile follows from the gradient, where a vertical reaction force of R acts downwards at the built-in end to restrain any upwards movement induced by the load. No external force is applied to the pin, there is no change in shear force, and no change in moment gradient there. The internal support reaction is $2R$ upwards – from the profile or simply by balancing the vertical forces knowing the built-in reaction.

Virtually the same propped cantilever is shown in Fig. 7.6(b), retaining the geometry, built-in end and loading as before. The pin-joint, however, has been

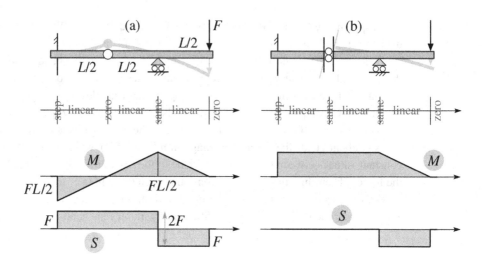

Figure 7.6 (a) Propped cantilever with internal pinned support. (b) Another propped cantilever with a sliding internal joint. Expected bending moment variations follow, with actual profiles, then shear forces.

replaced with a sliding joint, which now maintains the same rotation and, thus, bending moment across it.

The relative sliding is frictionless, and no shear force can be transferred across the joint. The axial separation, on the other hand, is fixed, enabling axial forces to be transmitted, if required, which we can ignore for the present loading.

In fact, there can be no shear force up to the internal support: none is certain in the first beam, which is in pure bending, and none in the first half of the second beam because no other forces are applied. After the internal support, where the bending moment is also $F \cdot (L/2)$, it drops linearly to zero; the shear force diagram follows trivially.

The differences between both sets of diagrams, between each structure, are quite stark. The function of the internal connection swaps around: from transmitting shear but no moment, to the opposite. Consequently, there is a more consistent bending moment profile for the second case but much less shear force compared to the first, and *vice versa*.

This change of one connection elicits conflicting performances throughout – think of how the internal beam bends in opposite directions despite the same downwards tip loading. Understandably, transmitting forces and moments through connections is as important as the member capacities elsewhere.

7.5   Moments in a Frame

Vertical beams become columns when part of a larger framework. A pair of columns joined by a horizontal top-beam creates a simple *portal* frame surrounding an

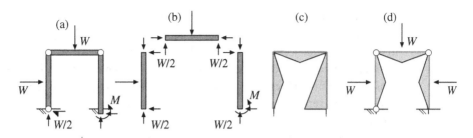

Figure 7.7 (a) Loaded, determinate portal frame. (b) Beam and column free bodies in equilibrium. (c) Corresponding bending moment variation plotted on the tensile side. (d) Extra horizontal W is applied right-side, giving a symmetrical set of moments.

open enclosure. Rows of portal frames are found in low-lying industrial buildings everywhere because they are efficient to build, structurally robust and relatively cheap.

A square portal frame of side-length L is shown in Fig. 7.7(a). The columns are pinned to the top-beam, and one of them is pinned to the ground; the other is built in at the base where a moment reaction can prevail. If, instead, we had a fourth pin-joint, the frame becomes a three-bar mechanism with synchronised large rotations of the beams incapable of resisting loads. By removing the pin and building in the column end, we constrain this motion most straightforwardly, giving us a statically determinate frame. Formal assessment of frame indeterminacy is given in the next chapter.

A single pair of horizontal and vertical forces, W, are applied to the mid-points of the left pinned column and the beam, as shown. The reaction force components are all $W/2$ but not inferred from global equilibrium: we may only say they must sum to W in either direction. The right approach divides the frame into its three constituent beams, Fig. 7.7(b).

The left-side column can only carry symmetrical forces because of its pinned ends (and because W is applied mid-way). The lateral forces are, thus, $W/2$, as indicated, which is the same shear reaction force at ground next to the base pin.

The top-beam also behaves symmetrically with vertical shear forces of $W/2$ at each end. Both of them are applied as axial forces to each column; and these cause vertical reaction forces of $W/2$ at their bases. The lateral force at the top of the left column is applied equally and oppositely to the beam, and through it to the right column. Considering its free body, the bending moment at the base must be equal to $(W/2) \cdot L$.

Because of the discrete loads and reactions, we expect linear bending moment profiles throughout. Other salient values are straightforward, namely $(W/2) \cdot (L/2)$ in the middle, and zero at all pins. But how are these profiles plotted? It is obvious for the top-beam because it is horizontal and behaves as if simply supported, and in appending its profile *to the beam*, as shown in Fig. 7.7(c), to the underside of it, we inadvertently plot this on the tensile side. If we adopt this convention throughout, there can be no ambiguity about bending moment directions in each of the columns.

We finish off with the same frame carrying an extra horizontal load, W, on the other column. Its presence may be expected to induce wholesale changes, tempting us to repeat our sequence of free-body analysis as before.

But the forces applied to the left column and top-beam cannot be different when we consider their free bodies. Furthermore, the loads applied overall are symmetrical with respect to the vertical centre-line of the frame; their effect must be reflected equally about this line. Formal details on symmetry are given in Chapter 9.

The new bending moment in the right column must be the same as the old unchanged bending moment in the left, which is drawn accordingly. Despite the built-in end, no moment is now applied there. If we wanted to, we may confirm the reactions at the base via shear force diagrams.

7.6   Final Remarks

Bending moment and shear force diagrams can be confidently tackled using the few simple rules given here. The sign convention adopted must be strictly followed, which also sets the signs within the governing equilibrium equations. The latter tell us what to expect of M and S according to the variation in w; crucially, we should draw bending moment profiles first before they are differentiated for shear.

A span without applied loads within a loaded frame or a continuous beam always experiences a linear bending moment along its length, and any discontinuous jump in M or S signifies an externally applied quantity – a couple (or torque), reaction force *etc.* By analogy with loaded cables, bending moments profiles are equivalent to lines, or curves, of tension under the given loads, but we must be careful about interpreting how the equivalent loaded beam is supported.

8   How Many Redundancies in Frames?

Maxwell's Rule for trusses from Chapter 4 expresses the level of indeterminacy from counting structural features. Any surplus of statical quantities become our redundancies and can be bar forces or support reactions. We identify them to complete our elastic analysis, which must invoke geometrical compatibility. Elsewhere, when applying, for example, Virtual Work or the Lower Bound Theorem, indeterminacy can give an economy of analysis as we have seen earlier.

Without changing the number of pin-joints, indeterminacy increases if more bars or support constraints are added as a thought experiment, and *vice versa* during their removal; removing too many, however, can lead to a mechanism. Determinacy is, thus, finely balanced between the catastrophic and the frustrated.

Frames and multi-span beams, like trusses, are a collection of interconnected members, where indeterminacy can also be couched in discrete terms. Unlike trusses, however, the connections between members and how they may be supported are more varied, and no straightforward rule for assessing indeterminacy naturally emerges.

Instead, we consider how the structure behaves when we alter the supports or member connections. They kinematically constrain the members but they can accord relative motions or *freedoms*, *e.g.* a roller-support imbues this duality most simply in the orthogonal: there is horizontal (say) freedom but vertical displacements are curtailed.

Increasing further the level of constraint introduces extra internal forces or moments, which raises the level of indeterminacy, whilst the opposite occurs when existing constraints are diminished. By focussing on these features specifically, we do not need to change the content level of the structure by removing or adding members themselves – as we might have done with bars in trusses.

The simplest kinematical connection is a pin-joint, which accords a relative rotation but no bending moment. Three bars in Fig. 8.1(a) are connected this way and to the ground by four pin-joints to give a familiar mechanism. Under virtually no loading, the arrangement 'sways' before falling over. The loading type is not important because mechanistic behaviour is a property of how the structure is built, and the load merely expedites collapse.

The structure is a truss mechanism, but it becomes a statically determinate frame in Fig. 8.1(b) when one of the grounded pin-joints is replaced by an *encastre* support, where its equilibrium was examined in Fig. 7.7. Free rotation relative to the base is now swapped for no rotation and a non-zero bending moment under loading.

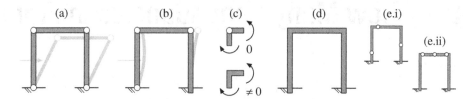

Figure 8.1 (a) Three-bar mechanism with four pin-joints. (b) Bottom-right pin from (a) is built in, to give a statically determinate portal frame. (c) Pin-joint allows for rotation between members but no bending moment (top); when replaced by a rigid joint, there is no relative movement and moments either side can be non-zero. (d) Indeterminate frame without pin-joints has three redundancies. (e.i) Reinstating three pins generally makes it determinate again; (e.ii) but not if the pins are locally arranged in the top-beam.

But in becoming determinate, we do not have to think about the specific balance between the number of equilibrium equations and all statical quantities. Rather, compared to (a), rigid-body rotation is clearly inhibited by the modified support and its extra bending moment, enabling the frame to become *simply stiff*.

Making any pin-joint rigid introduces a non-zero bending moment at that point, Fig. 8.1(c), and raises the indeterminacy level by unity. Alternatively, there is one fewer equilibrium equation because we can no longer excise a free body with that pin on the boundary of the body for which there is zero bending moment. Removing all pins to leave rigid joints throughout in Fig. 8.1(d) therefore gives us a frame with three redundancies.

Suffice to say, reinstating pin-joints reduces the number of redundancies because we give back locations with zero bending moment and, thus, more free bodies for which moment equilibrium is zero. These can be reinstated *anywhere*, not just at the original corners, to effect the same change in indeterminacy level. The positions in Fig 8.1(e.i) are viable but not so in (e.ii) because they all belong to the top member, which could *locally* collapse when loaded; how the rest of the frame responds remains unclear and so levels of indeterminacy are not clarified.

8.1   A Pin-Joint or a Simple Support?

Other kinematical modifications are possible, for example, adding a roller support to the freely-hinged rod in Fig. 8.2(a) returns a simply-stiff beam in (b). Rotations at the original pin-joint remain and, because displacements are still prevented, it is equivalent to a simple support, Fig. 8.2(c).

Replacing the pin with a built-in base produces a propped cantilever with one redundancy, Fig. 8.2(d), which can be the tip support reaction or any of the base reactions – one of them will always be 'spare' from equilibrium. Removing the tip support and hence its vertical reaction takes us back to a simple determinate cantilever in Fig. 8.2(e).

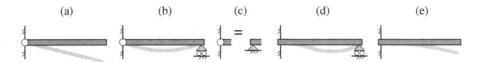

Figure 8.2 (a) Mechanistic cantilever. (b) Adding a roller tip support turns it into a simply-supported beam. (c) Equivalence of pinned and simple supports. (d) Indeterminate propped cantilever. (e) Now made determinate after removing tip support.

Figure 8.3 (a) Indeterminate beam built in at both ends. (b) Inserting a pin left and making the right end a roller support produces a determinate beam. (c) Cutting across the beam in (a) exposes three unknown stress resultants.

For practical structures, we have to work backwards from indeterminacy by inserting pins or otherwise until we reach obvious determinacy. For example, the beam in Fig. 8.3(a) is built-in at both ends. We are motivated towards a simply-supported beam by inserting pins at both ends: but one end must also be free horizontally, in order to mimic a roller-support, Fig. 8.3(b). Our original beam, therefore, has three redundancies.

We can also excise two cantilever free bodies by cutting across the beam at some point. In the same way that inserting a pin-joint eliminates a bending moment, cutting destroys three quantities that would otherwise exist: an axial force T, a shear force S and a bending moment M, Fig. 8.3(c). Each cantilever portion shares the same three quantities, so their relative lengths do no matter, and we confirm three original redundancies again. In destroying any statical quantity, we have *released* that quantity from the structure.

A slightly different approach is to build up to our final structure from determinate sub-structures, which are then connected together to reflect the actual layout. Obviously, any transmission of stress resultants from one sub-structure to the next are manifest as equal and opposite forces and moments acting on the individual sub-structure free-bodies; their tally then reflects the number of redundancies – after careful consideration.

For example, the tips of two cantilevers in Fig. 8.4 are re-connected via a pin-joint to give a right-angled frame with a free corner. A pair of orthogonal tip forces is thus applied to each cantilever, Fig. 8.4(a), despite not knowing what loading is applied; their built-in base reactions do not matter because we can calculate them from the tip loads, and there is no tip moment because of the prospective pin.

The tip loads are transmitted as shear forces, S_1 and S_2, and axial forces, T_1 and T_2. It seems there are four unknowns; however, these forces are locally applied to an elemental free body encapsulating the pin-joint, see Fig. 8.4(b). Its equilibrium

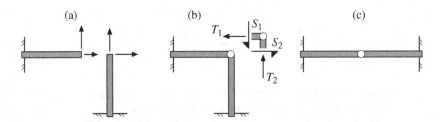

Figure 8.4 (a) Pair of cantilevers with unknown tip forces. (b) Now connected together by a pin-joint. Inset shows equilibrium of forces around an element of the joint. (c) Another pair of horizontal cantilevers connected by a pin.

from resolving orthogonally sets $S_1 = T_2$ and $S_2 = T_1$, reducing the total number of unknown forces to two, our two redundancies.

The horizontal cantilever pair in Fig. 8.4(c) is similarly governed. Their connectivity is the same as (b) if we imagine the original vertical cantilever being rotated by 90° anti-clockwise before connecting both halves together. Although the layout and geometry has changed, the number of structural features and their operation has not, which preserves the level of indeterminacy compared to (a). Note that if there were no pin, we have the beam in Fig. 8.3(a) again and its two cantilever halves expressing three redundancies: reinstating the actual pin takes us back to two.

8.2   Release the Trees

Cantilever sub-structures can, therefore, simplify our reckoning of indeterminacy. We have thought of them thus far as single member structures, of beams projected from a built-in support, horizontally or otherwise. They can be described more generally, starting off with the simplest variant of a bent cantilever in Fig. 8.5(a), with some nominal tip loading – forces, for convenience. As noted, any loading does not affect the degree of indeterminacy, but here it shows how actual stress resultants perform.

Isolating the horizontal portion as a free body, the tension T, shear force S and bending moment M at its base are found from equilibrium straightforwardly whatever the loading. These are reacted equally and oppositely onto the vertical cantilever stem in Fig. 8.5(b), which is also determinate, and the entire arrangement is, therefore, soluble.

Figure 8.5(c.i) shows a seemingly more complex example. However, we can subdivide this structure into separate cantilever portions as per (b) of any general orientation where, given the loading on each part, we find the forces and bending moments transmitted progressively inwards, to the final upright cantilever. The entire arrangement is also statical determinate, and notice how in (c.i) it resembles a *tree* structure.

The tree is characterised by off-shoot structures that do not reconnect to each other but form, in their own right, other cantilever sub-structures: no matter the level of recursive branching, our fundamental structure will always be determinate. If we can find equivalent sub-structures as free-bodies in our original structure, we know these

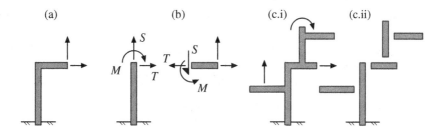

Figure 8.5 (a) Bent, right-angled cantilever. (b) Stress resultants at the corner junction. (c.i) General cantilever tree structure, where no branches reconnect; (c.ii) sub-division into straight, determinate cantilever portions.

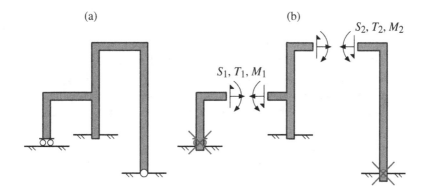

Figure 8.6 (a) Double and stepped portal frame with roller, built-in and pinned supports. (b) Cantilever trees from (a) by building in all supports and cutting across the two beams.

to be determinate, and for each cut we make to achieve these separations we tally three more redundancies. Let us explore this approach in three final examples.

Figure 8.6(a) indicates a double portal frame whose ground positions are stepped. There are three vertical columns and two horizontal beams, and one grounded pin-joint. At the left-side foot there is a horizontal roller support which prevents rotations.

Any cantilever tree must be built-in at its base, so the pin and roller support are temporarily 'scored out' and replaced with encastre supports in Fig. 8.6(b). These react a new bending moment and horizontal force, respectively, that must be discounted from the total number of redundancies. Both beams are cut across to reveal three cantilever trees with interrelated pairs of stress resultants S_1, T_1, M_1 and S_2, T_2, M_2. The total number of redundancies is, therefore, six minus two, *i.e.* four.

The second example is a ring – a self-connecting circular beam. It is not supported in Fig. 8.7(a) and will accelerate out of view under loading. So first, we cut across the ring then clamp one of the exposed ends, Fig. 8.7(b). We now have a circular cantilever, about to reconnect to itself but not quite; the cut accommodates two forces and a moment (not shown), which, if known, make the system determinate: but we presume they are not known and so we have three redundancies.

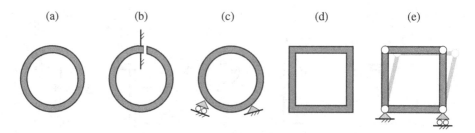

Figure 8.7 (a) Ring beam. (b) Now cut then clamped to reveal a curved cantilever. (c) Instead of clamping the ring, pin- and roller supports provide overall rigid-body restraint. (d) Square frame. (e) Now supported and with pins inserted, to yield a four-bar mechanism.

Clamping the loaded ring itself prevents any rigid-body motion. This removes, at a point, any linear motion in two directions and a rotation. The same can be achieved using simple supports instead; in Fig. 8.7(c), the fixed support, which prevents planar translations, couples to the roller to curtail overall rotation.

The square frame in Fig. 8.7(d) behaves identically even though it has a different geometry. We can repeat the process of cutting across and clamping an exposed end, resulting in the same three redundancies, but we can also introduce pin-joints until the frame becomes a mechanistic truss provided we also support it to suppress other rigid-body motions.

Inserting four corner pins results in a four-bar (or three-bar, if the ground is not considered to be equivalent to a fixed bar) mechanism in Fig. 8.7(e), with the same supports as the ring in (c), where $b + r - D \cdot j = 4 + 3 - 2 \times 4 = -1 = -m$ from Maxwell's Rule in Chapter 4. Inserting three pins, *i.e.* one less, adds unity to this result and brings us back to determinacy.

The third example is another double portal frame with a pin-joint over the middle column, Fig. 8.8(a). Following our previous approach, we remove the joint and find three trees from two cuts, Fig. 8.8(b), giving six minus one, *i.e.* five redundancies.

There are, in fact, four redundancies. The pin-joint connects *three* separate portions of the frame, the column and two beams either side. It therefore permits relative rotations between all of them, or, if we fix one of the portions, the other two can rotate independently. The pin at the junction must, therefore, accord two kinematical freedoms, Fig. 8.8(c), not the usual one. The same is achieved by having two pin-joints separated just on either side of the junction: we can imagine the current pin-joint being formed by these two pins then coalescing at the junction, Fig. 8.8(d).

8.3   Final Remarks

Statical indeterminacy in beams and frames is also a property of their layout and support conditions, and nothing to do with the loading, which only makes the redundancies larger or smaller in size. We choose what becomes a redundancy – for mathematical convenience in whatever analysis follows; the structure does not 'decide.'

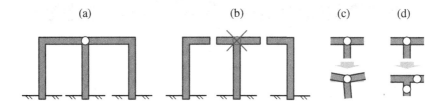

Figure 8.8 (a) Double portal frame with central pin-joint. (b) Equivalent cantilever trees after removing the pin. (c) Central pin connects three elements with two relative rotations. (d) Two pins placed very close to the central junction accord the same kinematical freedoms as the central pin alone.

Producing a kinematical freedom is tantamount to destroying a stress resultant or support restraint. We do this by cutting, inserting pins or making cantilever trees, being careful at frame junctions and ground supports. Each freedom brings us closer to determinacy, and sometimes it is better to go beyond by one freedom in order to identify an obvious single degree-of-freedom mechanism.

Understanding indeterminacy by reassembling and reconnecting determinate substructures is another approach. Every kinematical constraint added raises the redundancy level by unity, but any original freedoms in the structure (such as pins or roller supports) may have to be deleted initially before reinstating at the end (and then subtracting from the total number of redundancies).

9 Symmetry and Anti-symmetry

What can we say about the *nature* of the elastic joint displacements in Fig. 9.1(a) when a horizontal force, F, is applied to the right-angled, pin-jointed arch? Specifically: what zero values, if any, can be asserted for either of its orthogonal components, to simplify any calculation later?

The lateral component, δ_H, cannot be zero because the joint is free to deflect in the same direction as F. The performance of the vertical component, δ_V, is less obvious, but let us conduct a thought experiment. Imagine that F is applied instead in the opposite direction, Fig. 9.1(b). We can calculate the bar tensions again from nodal equilibrium but they should only change their signs compared to before. Our displacements should also change sign accordingly.

However, we have deliberately chosen a structure with a single plane of vertical symmetry through the loaded joint and normal to the plane of the page. This plane is the dashed line in Fig. 9.1(b) about which the structural layout is perfectly reflected; compared to Fig. 9.1(a), F has also been reflected, as are the drawn displacements in (b). The horizontal component switches direction; however, the vertical component is unaltered because it is parallel to the symmetry line.

We now have a conflict of performance for δ_V: it ought to change sign if F is reversed, but this flouts the symmetry expectation. Satisfaction throughout arises only if δ_V is *zero*. Alternatively, we can *superpose* the loadings from Figs. 9.1(a) and (b) *after* asserting our symmetry expectations to give the structure in Fig. 9.1(c), which is clearly unloaded. Superposition demands that we must add the displacements too, where we are left with $2\delta_V$. This is impossible for no loads applied, and δ_V must be zero again.

In moving from Fig. 9.1(a) to (b) we have, equivalently, *flipped* the structure, by 'lifting' it and its vector quantities out of the plane of the page and turning them over about a vertical axis before reinstating in (b). We have not changed the structure, its loading or its deflection response: we are merely looking at (a) through the underside of the page.

Because our original loading changes direction if we impose reflection about the symmetry line, we say that it is *anti-symmetrical*, along with δ_H but not δ_V. More generally, symmetrical structures loaded anti-symmetrically experience only anti-symmetrical displacements (and rotations), as we shall see in other examples; other 'combinations' of rules are given later.

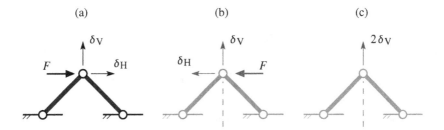

Figure 9.1 (a) Symmetrical pin-jointed truss loaded horizontally. (b) Same truss with the loading reversed: the dashed line is a vertical plane of layout symmetry. (c) Superposition of the loadings and displacement components from cases (a) and (b).

9.1 The Needless Bar

The pin-jointed framework in Fig. 9.2(a) is similar to before, but three bars are now connected to a single loading joint, giving statical indeterminacy. Alternatively, and formally, from Maxwell's Rule in Chapter 4: $b+r-D\cdot j = 3+6-2\times4 = 1 = s$. But the structure is again symmetrical about the vertical bar, and we can use the previous arguments to clarify the relationships between the reaction force components.

First, we ascribe general reaction force components, horizontally and vertically, at each support in Fig. 9.2(b). For the first grounded joint, these are H_1 and V_1 of arbitrary positive directions. We anticipate a single vertical component for the middle grounded joint because the connected bar is vertical; it must be equal to the tension in the bar, T_2. The final joint has components H_3 and V_3 as usual.

We flip the structure about a vertical axis as before to give the viewpoint in Fig. 9.2(c). Apart from T_2, the reactions have exchanged positions, H_1 and H_3 and the load have also changed directions: since T_2 lies on the symmetry plane, it is unchanged. The geometry of the structure is unchanged compared to Fig. 9.2(b), and we may superpose both in order to give a structure which is unloaded. Consequently, the reactions must be zero, from which $H_1 = H_3$ and $V_1 = -V_3$, and $T_2 = 0$ (following $2V_2 = 0$).

This final result is unexpected: a formal analysis would give the same result only after significant working. But we see immediately a soluble truss: setting the other bar tensions to be T_1 and T_3 in Fig. 9.2(d), we find from equilibrium of the loaded joint that $T_1 = -T_3 = F/\sqrt{2}$. These tensions also confirm $H_1 = H_3 = F/2$ and $V_1 = -V_3 = F/2$ from equilibrium of the grounded joints: the benefits of exploring symmetry are clear.

9.2 Flipping

Flipping and superposing a symmetrical structure with itself reduces unknown information and simplifies any analysis which follows. These procedures can be extended

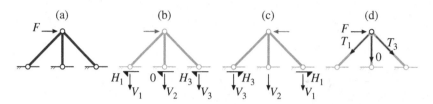

Figure 9.2 (a) Symmetrical indeterminate truss loaded horizontally. (b) Reaction force components at the supports. (c) Reversal of loadings and reactions. (d) Final set of bar forces.

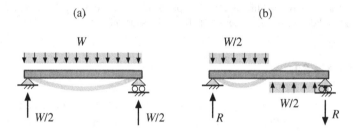

Figure 9.3 (a) Uniformly loaded beam, responding symmetrically. (b) Same beam loaded anti-symmetrically, responding the same.

to finding out about internal properties on the symmetry line when we operate upon one half of the structure as a free body.

In Fig. 9.3(a), the uniformly loaded, simply-supported beam has symmetrical vertical reactions of $W/2$. The roller support on the right side can be replaced by a fixed pinned support, as per left, for a perfectly symmetrical determinate case only because no horizontal forces are applied. The central stress resultants for this are well known: the bending moment is a maximum and the shear force is zero. The symmetrical displacements also tell us that the central rotation is zero.

Figure 9.3(b) shows the same beam now loaded anti-symmetrically with a total load on each half of $W/2$. The load gives no net vertical force, so the (unknown) reactions, R, are equal and opposite: R cannot be zero – another viable equilibrium solution from taking moments about either support. The displacement profile has been assumed anti-symmetrical also, giving a zero value in the middle: we now verify as such, amongst other performances.

Even though the loading is anti-symmetrical, we consider one-half of the initial symmetrical beam either side of the central symmetry plane. Choosing the left-hand side in Fig. 9.4(a), nothing is assumed about the central quantities shear force S, moment M, displacement δ and rotation θ. We do, however, assume a positive direction for each of them, as drawn; and these must be consistent with the applied loading and reaction directions.

The half-beam is then flipped over in Fig. 9.4(b) where the directions of M and θ are reversed: this is our back-side view of (a). If we then multiply all quantities by minus one, we effectively invert behaviour, Fig. 9.4(c): we also obtain something

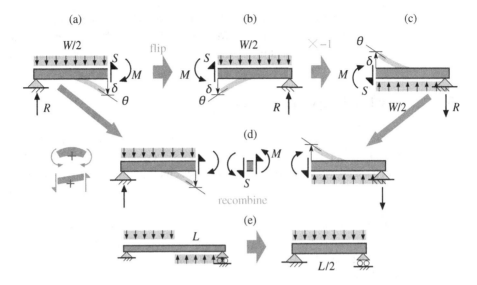

Figure 9.4 (a) Left half of the beam from Fig. 9.3(b) with non-zero centre-line quantities. (b) Now flipped compared to (a), with all positive directions of quantities preserved. (c) Reversal of loading and response of case (b). (d) Halves from (a) and (c) are recombined to reveal the elemental equilibrium performance on the original beam centre-line. (e) Equivalent half-beam for calculating detailed performance along the beam.

resembling the right-half of Fig. 9.3(b) and how it is loaded with the directions of centre-line quantities being consistent with their positive definitions in Fig. 9.4(a).

We can bring together, or 'recombine', the two halves in Fig. 9.4(d) as free bodies of our original loaded beam. Between them, we can imagine an infinitesimal central element carrying only shear forces and bending moments but no loading in the limit of zero width.

The transfer of shear and moment to each side from each half must be equal and opposite, as shown. Whilst no net shear force is applied to the element, it carries a net anti-clockwise moment of $2M$. This violates equilibrium and, thus, M must be zero; the shear force, however, does not and is non-zero.

Even though the central displacements are exaggerated for each half, they jump from the under-side to the top-side of the beam across the middle. Again, this is impossible, so δ must be zero as we proposed initially. The rotations, on the other hand, are aligned without a discontinuity in gradient and can be non-zero. Bringing the displaced points together whilst maintaining the same relative rotations would achieve the anti-symmetrical profile assumed initially in Fig. 9.3(b).

We now confidently understand how both halves of the structure work, specifically, the boundary conditions if we consider either half by itself. We have a simpler *analogue* of the original symmetrically loaded beam, which commands less working, re-drawn in Fig. 9.4(e).

It has two simple supports: the original left-side one and the right-side one corresponding to zero displacement in the middle of the original beam. The span is $L/2$, and it is loaded uniformly and only by $W/2$: no right-side end moment is applied because M is zero in the original middle. We can now find, say, the maximum deflection, end rotation, bending moment profile *etc.* by techniques delivered in Chapter 11.

9.3   Equal and Opposite Reactions

The square portal frame in Fig. 9.5(a) is a symmetrical beam and column structure with built-in feet. It has three redundancies according to the method of Chapter 8, which we take to be the base reactions. We actually show six different components, H_1 *etc.*, because we do not know how they are related to each other. A central horizontal force, W, is applied.

Taking moments and resolving in two directions enables the six unknown reactions to become three, but we still cannot apportion specific information. For example, it is tempting to say that H_1 and H_2 share W *equally*; resolving horizontally only suggests that H_1 and H_2 *sum* to W, nothing else. Resolving vertically tells us that $V_1 = -V_2$ for certain, but when we take moments there is similar uncertainty about the specificity of the moments M_1 and M_2.

In each separate view of the reactions, Figs. 9.5(b)–(d), we flip the structure and reproduce the back-side views below. These are then added together in the final row to yield an unloaded structure because W is applied on the symmetry line. Resolving vertically in Fig. 9.5(b), the reactions at either support confirm $V_1 + V_2 = 0$ *i.e.* $V_1 = -V_2$ equal to V, say.

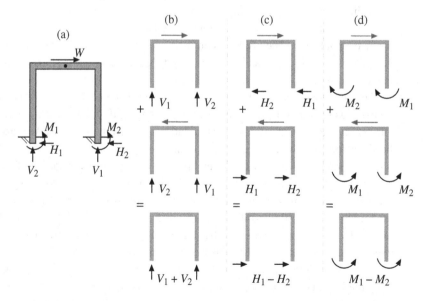

Figure 9.5 (a) Indeterminate portal frame loaded in the centre of the top beam (black dot) and horizontally, with general reaction forces and moments. (b)–(d) Top row: each set of original force components and moments; middle row: flipped views of the top row; bottom row: superposed loadings from the top two rows.

At each foot in Fig. 9.5(c), H_1 must be equal to H_2 because $H_1 - H_2 = 0$, each being equal to $W/2$ as we surmised. Figure 9.5(d) indicates that $M_1 = M_2 (= M)$, and taking moments about any point in the original frame, say the left foot, we have: $2M - WL - VL = 0$ for a side-length of L. Thus, M is equal to $(W + V)L/2$ where V is now our sole redundancy when we started with three.

9.4 Spinning, not Flipping

Our superposing can also be extended to structures with anti-symmetrical layouts. Their geometry is preserved by *rotating* 180° about a central origin; alternatively, we may think of reversing one half of the structure after it is reflected about a central symmetry plane.

The *stepped* portal frame in Fig. 9.6(a) is an example, and is loaded centrally by a vertical force F. It is indeterminate – the degree does not matter at this stage, and its six general reactions are indicated with two forces and one moment at each foot: the height and width are both L. Our aim is to relate F to the reactions.

Obviously $H_1 = H_2$, equal to H, from resolving horizontally. However, this says nothing about their absolute values, so we keep H_1 and H_2 in what follows. First, the structure is rotated about its centre by 180° to give an 'unchanged' geometry in Fig. 9.6(b). The various reactions have exchanged positions and, some, also their directions, but the statical character is the same.

The loadings from (a) and (b) are now superposed in Fig. 9.6(c) to give no loading throughout. Immediately $H_1 = -H_2$, but this contradicts our earlier statement, and both must be zero. We also find $V_1 = -V_2$, equal to V, giving the same direction for both in Fig. 9.6(a) and $V = F/2$: from the moment reaction, $M_1 = M_2 = M$.

Our found reactions are described in Fig. 9.6(d), which confirms overall moment equilibrium (and that H must be zero accordingly). Our single redundant reaction is therefore M when ordinarily we might expect three for this type of frame.

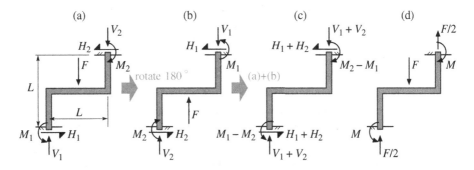

Figure 9.6 (a) Stepped portal frame loaded centrally and vertically, and its general reaction force components and moments. (b) Case from (a) is rotated by 180°, preserving all positive directions. (c) Cases (a) and (b) superposed to give an unloaded frame. (d) Final set of non-zero reactions for the original loading.

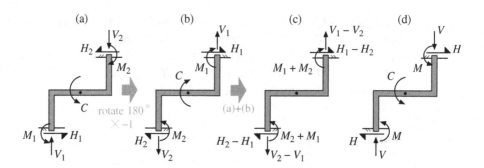

Figure 9.7 (a) Stepped frame from Fig. 9.6 now carries a central (black dot) couple. (b) Case (a) rotated by 180°. (c) Then superposed with (a) for no loading. (d) Final set of reactions where $M = [C + (H - V)L]/2$.

The same frame now carries a central couple C, Fig. 9.7(a), with again six general reaction components. We rotate the frame, multiply all statical quantities by -1, Fig. 9.7(b), then superpose with (a) in Fig. 9.7(c) to give the unloaded structure.

From the zero reactions, $H_1 = H_2 = H$, $V_1 = V_2 = V$ and $M_1 = -M_2 = M$. Seemingly we have three redundancies, but overall moment equilibrium in Fig. 9.7(a), after substitution, returns an extra relationship between them: $2M + C + (H - V)L = 0$. We now have two, fully described in Fig. 9.7(d).

As a final question, we consider the symmetry state of these latest results. For symmetrical structures we noted that its response was preserved with respect to the symmetry plane whilst anti-symmetrical behaviour reverses. In Fig. 9.6(d), the reactions behave symmetrically by these measures even though the underlying layout is anti-symmetrical. Alternatively, we can rotate one half of the structure by 180° before reversing the reactions to give the same. Our original loading is symmetrical, so we see a symmetrical response for an anti-symmetrical layout. On the other hand, C and its found reactions are anti-symmetrical in Fig. 9.7.

The corresponding displacement profiles are given in Fig. 9.8 whose symmetry states are, however, not obviously expressed. For loading F, the horizontal beam deflects symmetrically with equal and opposite rotations at its jointed ends which, if rigid, dictate the displaced profiles in each of the vertical parts, Fig. 9.8(a.i). The latter do not simply reflect each other in the symmetry plane because they belong to an underlying anti-symmetrical layout.

Instead, the left half in Fig. 9.8(b.i) is rotated by 180° in Fig. 9.8(c.i) before reversing its displacements in Fig. 9.8(d.i). If the two halves are combined, (b.i) and (d.i), we have the original displacements on the anti-symmetrical layout. The transformation is the same as per the reactions, so the displacements can be declared symmetrical.

The displacements from the couple C, Fig. 9.8(a.ii), are surmised similarly from first the horizontal beam, its end rotations and the rigid joints, and simply rotating one half of the structure by 180° in Figs. 9.8(b.ii) and (c.ii) confirms their anti-symmetrical nature.

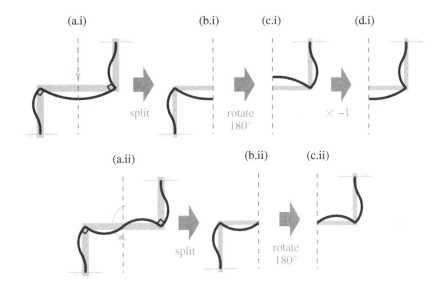

Figure 9.8 Top row, (a.i–d.i): symmetrical determination of the displacements from Fig. 9.6. Bottom row, (a.ii–c.ii): anti-symmetrical determination of the displacements from Fig. 9.7.

9.5   Final Remarks

Symmetry analysis can be approached formally using the mathematics of Group Theory. We do not have to because our structures are relatively simple with a single plane of symmetry (or anti-symmetry), and flipping, or spinning, is our *ad hoc* method for establishing the symmetry character of properties on this line.

We can also confirm that the symmetry character of moments and shear forces on this line obeys a simpler rule. From equilibrium of an element about this line, non-zero moments must be symmetrical (as are axial forces); equal and opposite shear forces are anti-symmetrical by their nature.

Displacements are similarly symmetrical but rotations are not, and with confidence we can begin to ascribe non-zero (and zero) quantities by inspection; if there is a doubt, proceed with flipping *etc*. A reduced model, typically involving half of the structure, will always be the outcome with appropriate boundary conditions, either kinematical or statical, on the (anti-)symmetry line.

Part IV

Beams and Frames: Analysis

10 Standard Beam Deflections

So far in beams, we have dealt largely with their equilibrium performance: of the relationship between bending and shear, and transverse loading, and of how to draw profiles of the former skilfully and efficiently, complemented as ever by informal sketches of the displaced shape.

These displacements *have to be* small, for two reasons. Naturally small deflections remain elastic and give an overall high stiffness, where visible movements of the structure are negligible. Consequently, statical propriety is assured and unwanted dynamical vibration is also limited, which extends the fatigue life of the structure.

Mathematically, they also allow us to write equilibrium statements with respect to the original geometry of the structure. Not only does this simplify their expressions but it uncouples the kinematics from equilibrium during analysis. The analytical steps involved now are executed in series rather than in parallel, making the analysis simpler again.

Calculating displacements, therefore, tells us of preliminary design information, which we do first for *standard* beam cases. These are determinate cantilevers and simply-supported, single span beams, with straightforward point loading or uniform loading intensities. Our aim is not in the method but in the final results: later, we focus on the methodology for general cases, including indeterminate ones.

10.1 Method

Recall our three imperatives of equilibrium, geometrical compatibility and the generalised Hooke's Law. These are associated to how an element of the structure behaves, and connect exterior quantities to interior ones. Later, in Chapter 13, we shall see how to choose the correct, or *dominant*, structural action when there is a 'choice' between beam bending or bar behaviour.

The equilibrium statement was given originally in Eq. (7.1), which we re-state with respect to a right-ward moving beam coordinate, x:

$$\frac{dS}{dx} = w \quad \text{(a)}, \quad \frac{dM}{dx} = S \quad \text{(b)}; \quad \frac{d^2M}{dx^2} = w \quad \text{(c)}. \tag{10.1}$$

Across a given beam element in the direction of x, positive moments are hogging, w acts downwards and S swaps from down to up.

The compatibility relationship is between the transverse displacement, v, and curvature, κ, of the beam neutral axis. Recall from the Author's note that the latter is the rate of change of tangent angle with respect to the beam axial coordinate. This angle and shallow gradients are one in the same when displacements are small, where Eq. (0.5) between κ and v follows.

Henceforth we shall introduce a minus sign to its right-hand side. If, as in the earlier example, we choose a descriptive positive quadratic for v, our beam centre-line curves upwards with negative curvature. Thus, we associate downwards curving with positive hogging behaviour, where

$$\kappa = -\frac{d^2v}{dx^2}. \tag{10.2}$$

Alternatively, we may think of the *centre* of positive curvature lying below the trajectory of the beam centre-line.

The generalised Hooke's Law for this problem relates bending moment and beam curvature linearly by the bending stiffness, EI, according to $M = EI\kappa$. Replacing κ with Eq. (10.2) and then M with Eq. (10.1)(c), we arrive at our final governing equation of deformation:

$$\frac{d^2}{dx^2}\left(EI\frac{d^2v}{dx^2}\right) = -w. \tag{10.3}$$

Note first that the EI term is nested inside before differentiating, which is correct if it can vary with x. Usually, it is constant (E certainly is) but if the cross-section shape changes or tapers along the beam, then I is a function of x. We have also moved the minus sign to the right-hand side for convenience.

We now have a linear differential equation in terms of exterior quantities – the transverse loading and displacements, which can be integrated exactly in simple enough cases to relate pertinent values. For example, consider the uniformly loaded, simply-supported beam in Fig. 10.1(a).

The loading intensity is w N/m, giving vertical reaction forces $wL/2$ at both ends of the span of length L. The general solution for Eq. (10.3) is now a fourth-order polynomial expression, $v = a_0 + a_1x + \cdots a_4x^4$, whose unknown coefficients, a_i, are stipulated by the boundary conditions, *i.e.* known geometrical features from its shape or structural behaviour at certain positions.

However, let us set the origin of x to be in the middle of the beam. We expect symmetrical displacements, and, thus, terms a_1x and a_3x^3 disappear because they are not, and $v = a_0 + a_2x^2 + a_4x^4$.

We now require three boundary conditions, which are furnished by the geometry of deformation and statics together. At each end, there is no moment and hence no curvature, which tell us about the second derivative of v:

$$\frac{d^2v}{dx^2}\Big|_{x=\pm L/2} = 0 = 2a_2 + 12a_4x^2 \quad \rightarrow \quad a_2 = -6a_4L^2. \tag{10.4}$$

There is also no deflection at both ends, which is satisfied simultaneously by $a_0 + a_2 L^2/4 + a_4 L^4/16 = 0$, giving the central deflection, $v_{x=0}$, equal to $a_0 = 5a_4 L^4/16$ after substituting for a_2 from Eq. (10.4).

Finally, the shear forces at both ends are captured using Eq. (10.1)(b) after replacing M and κ via Hooke's Law to return $S = -EId^3 v/dx^3$. The right-hand side is equal to $\pm 12EIa_4 L$ at $x = \pm L/2$, where the shear forces, are respectively, $\mp wL/2$, recalling our usual sign convention for S. We therefore find $a_4 = -w/24EI$ and

$$v|_{x=0} = -\frac{5wL^4}{384EI} \qquad (10.5)$$

where the minus sign signifies a downwards displacement.

We have, as noted, a relationship between the exterior, central deflection and the applied loading in terms of intrinsic properties of material stiffness (E) and geometry from its span (L) and cross-sectional proportions (I). This elegant, relatively simple expression is remarkable when we remember that we are capturing the response of an otherwise complex three-dimensional elastic system.

10.2 Another Example

If we write instead a ratio of load-to-deflection, w/v (or in total load, wL/v), we express by definition a stiffness for the structure itself, loaded in this way: beams, like bars, are uni-dimensional springs whatever the loading, even the tip-loaded cantilever in Fig. 10.1(b).

There are obvious differences, notably the built-in (recall *encastre*) base support, which expresses different boundary conditions for that end of the beam. Without any distributed loading, $w = 0$ in Eq. (10.3) and displacements are served by the complementary function only. Because of the fourth order variation, v must be a cubic polynomial where $v = b_1 x + b_2 x^2 + b_3 x^3$ for the origin of x at the base.

Symmetry does not apply in this case, but we see that the coefficient b_1 is zero immediately from having no gradient at the base where $dv/dx = 0$. Only two coefficients remain, found from two pieces of information.

First, the shear force at the base is $-F$ from the usual sign convention when the tip load, F, acts downwards. Differentiating v three times and setting equal to $-S/EI$, we find $b_3 = F/6EI$. The positive bending moment, FL, at $x = 0$ returns $b_2 = -FL/2EI$, and the variation of displacements is thus:

$$v = \frac{Fx^3}{6EI} - \frac{FLx^2}{2EI} \quad \rightarrow \quad v|_{x=L} = -\frac{FL^3}{3EI}. \qquad (10.6)$$

The tip rotation, θ, is set by dv/dx at $x = L$, i.e. $\theta = -FL^2/2EI$, dipping downwards under the minus sign.

A second cantilever in Fig. 10.1(c) is now loaded by an end couple, C. It is tempting to repeat our previous approaches, but we note that everywhere the bending moment

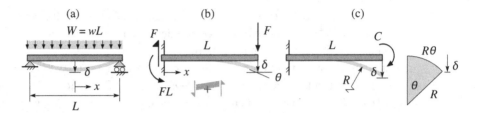

Figure 10.1 (a) Simply-supported beam with uniform loading intensity, $w = W/L$, and centre origin for x. (b) Tip-loaded cantilever by an external force. (c) Loaded by a couple, giving uniform curving.

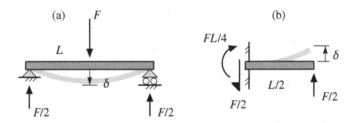

Figure 10.2 (a) Simply-supported beam loaded centrally by a point force. (b) Statically equivalent 'half'-cantilever, with the same end deflection as the central one in (a).

is constant, positive and equal to C. From Hooke's Law, the curvature everywhere behaves the same: $\kappa = C/EI$.

Its inverse is, therefore, a uniform radius of curvature R, redrawn in (c) as part of a circular sector of the same radius pertaining to the arc shape of the deflected cantilever neutral axis. Its arc-length is fixed at L and equal to $R\theta$, where θ is the tip rotation and the angle subtended along the arc from the base. Straightaway, $\theta = CL/EI$ when we replace R in $\theta = L/R$ from Hooke's Law.

The vertical tip deflection downwards is now $\delta = R(1 - \cos\theta)$, *exactly*. But if gradients are shallow then θ is small and $\cos\theta$ is approximately $1 - \theta^2/2$. Therefore $\delta \approx R\theta^2/2$, and substituting for θ and R as before, we can write down $\delta = CL^2/2EI$ with no more calculation.

The central deflection, δ, of the centrally loaded, simply-supported beam in Fig. 10.2(a) can be solved for indirectly from the earlier result. If we excise half of the symmetrically displaced beam to the right (or left) of the load, W, it has zero gradient adjacent to this. At the same position, the original free body has a bending moment, $FL/4$, along with shear forces $F/2$ at both ends. No other external loads exist and we have, in fact, the same forces and moment acting on a cantilever of the same length, $L/2$, Fig. 10.2(b), *i.e.* statical equivalency.

The tip deflection is upwards but crucially must be the same size as the downwards δ in (a). We can therefore use the cantilever result from Eq. (10.6) where L is replaced by $L/2$ and F by $F/2$, giving $\delta = FL^3/48EI$.

Table 10.1 SS signifies simply-supported: PL for point loading: UDL for uniformly distributed loading. All beams and cantilevers have a span or length equal to L, and the location of displacements and rotations are indicated.

beam type	displacement	rotation
SS, central PL of F	$FL^3/48EI$ (centre)	$FL^2/16EI$ (end)
SS, UDL of $W = wL$ (total)	$5WL^3/384EI$ (centre)	$WL^2/24EI$ (end)
cantilever, tip PL of F	$FL^3/3EI$ (end)	$FL^2/2EI$ (end)
cantilever, UDL of $W = wL$	$WL^3/8EI$ (end)	$WL^2/6EI$ (end)
cantilever, end moment of C	$CL^2/2EI$ (end)	CL/EI (end)

10.3 Final Remarks

The polynomial substitutions are examples of *shape functions*, which satisfy the general governing equation of deformation. In effect, we solve a differential equation by inspection only because the loading and support conditions are amenable to a polynomial description. If instead w was a trigonometric function, *e.g.* proportional to $\sin \pi x/L$, then so is the particular integral for v.

These loaded beams are ubiquitous and their responses should be committed to memory. The complete set, which includes end rotations, is given in Table 10.1 with an extra line for a uniformly loaded cantilever (deduced via the earlier simply-supported beam).

Other loading cases for the same types of beams can be solved individually. However, using statical equivalency and superposition together in the next chapter, we can recycle our standard results for an alternative, and simpler, solution; in effect, we turn a differential problem into a compilation of algebraic solutions.

This approach can be extended to non-standard beam layouts if their element sections correspond to standard cases. The boundary conditions between sections must reflect the connection type, *e.g.* a simple support or a rigid corner; and their geometrical compatibility produces extra, algebraic equations for a complete solution. Writing these equations explicitly in terms of displacements or rotations gives so-called *deflection coefficients*, which are equated for shared boundary conditions. Accordingly, we can tackle problems of indeterminate beams and frames.

11 Deflection Coefficients

The standard beam deflection cases from Table 10.1 in Chapter 10 speak to simply-loaded determinate cases. These can also be used to solve indeterminate structures loaded generally by thinking about how equilibrium and kinematics interact. First, let us consider a non-standard determinate case to emphasise the use of superposition in our proposed method.

11.1 Moments from the End

A moment C is applied to one end of a uniform simply-supported beam with bending stiffness EI in Fig. 11.1(a). The displacement variation will be non-uniform because of the loading asymmetry, but given the nature of the loading it is appropriate to deal with end rotations instead, both at C and the free end. These are denoted as θ_1 and θ_2, respectively, over a support span of L, which we now find.

First, we decouple the original loading into symmetrical and anti-symmetrical components, *c.f.* Chapter 9, which can be recombined by superposing, Fig. 11.1(b): one case has equal end moments, $C/2$, and the other with equal and opposite moments of the same.

Because the original beam is symmetrical about the centre, the end rotations are either symmetrical and denoted as θ_3, or anti-symmetrical, θ_4, from which $\theta_1 = \theta_3 + \theta_4$ and $\theta_2 = \theta_3 - \theta_4$ for our assumed directions of positive rotation. For now, statical equivalency tells us that θ_3 will also be the rotation at the end of a cantilever of half-length $L/2$ under $C/2$, Fig. 11.1(c.i), where Table 10.1 gives $\theta_3 = (C/2) \cdot (L/2)/EI = CL/4EI$.

Finding θ_4 is, however, more challenging. The central deflection in this case is clearly zero from anti-symmetry, as is the bending moment there: if we apply the flipping-superposition method from Chapter 9 to one half of the beam, these properties emerge straightforwardly. Thus, we can treat Fig. 11.1(b.ii) in terms of its half span – we choose the left side – simply supported at the right-side, which is the original middle of the beam: see Fig. 11.1(c.ii).

But recognise that Fig. 11.1(c.ii) is now statically *similar* to the original problem in Fig. 11.1(a), where the span and applied moment are now both halved. We therefore expect θ_4 to behave in proportion to θ_1, but it is not quite clear what that proportion is.

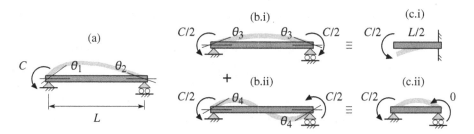

Figure 11.1 (a) Simply-supported beam loaded by end couple. (b.i) Symmetrical and (b.ii) anti-symmetrical loading components. (c.i) Equivalent cantilever and loading for finding the rotation in (b.i); (c.ii) equivalent simply-supported beam related to (b.ii).

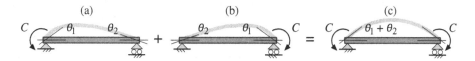

Figure 11.2 (a) Same beam from Fig. 11.1(a). (b) Now flipped about a vertical axis to give a back-side view. (c) Cases (a) and (b) superposed to give a symmetrically loaded beam.

We return briefly to Fig. 11.1(c.i). We know that θ_3 depends linearly on C and on L: if we double C, the rotation doubles *etc.* Such dependency also reinforces the *dimensional* expectation that θ_1 (and θ_2) in Fig. 11.1(a) are also related to the product of applied moment and span, for we have a linear elastic beam supported differently but loaded similarly to the cantilever.

More formally, we may write $\theta_1 = CL\beta$ where β is some constant of proportionality for this problem, which obviously includes EI: if C or L changes, so does θ_1 but not β. Consequently, θ_4 must be equal to $(C/2) \cdot (L/2)\beta$, *i.e.* $\theta_1/4$. We already know that $\theta_1 = \theta_3 + \theta_4$; replacing θ_4 with $\theta_1/4$, we find $\theta_1 = 4\theta_3/3$ or $\theta_1 = CL/3EI$.

Finally, finding θ_2 exploits superposition in Fig. 11.2. The original case, repeated in Fig. 11.2(a), is reversed in view in Fig. 11.2(b). These are added together in Fig. 11.2(c) to give a symmetrically loaded beam with equal end rotations $\theta_1 + \theta_2$. This must be the same rotation at the end of a half-length cantilever under the same end moment, $C(L/2)/EI$. Substituting for θ_1 from above, we find $\theta_2 = CL/6EI$ or that $\theta_2 = \theta_1/2$ in general.

11.2 Propping Up

The propped cantilever in Fig. 11.3(a) carries a uniformly distributed total load, W, the span is L, and the vertical reaction forces F and R along with base moment C are indicated.

The system has one redundant quantity, which we will ascribe to one of the reactions: the choice, as we shall see, does not affect the final result. Irrespective of that choice, we must also think about how compatibility affords extra information for a complete solution, in particular, the kinematic constraint commensurate with our redundancy choice.

We first choose the contact reaction force R at the tip. It arises because the support there prevents otherwise 'free' displacement of the tip from W. We can visualise this action as two separate loading processes; first, where there is no tip support and the displacement is δ_1, and second, when R is applied to the tip with no load, giving equal and opposite δ_2. The scenario is depicted in Fig. 11.3(b), where Table 10.1 assists:

$$\delta_1 = \frac{WL^3}{8EI}, \quad \delta_2 = \frac{RL^3}{3EI} : \quad \delta_1 = \delta_2 \quad \rightarrow \quad R = \frac{3W}{8}. \tag{11.1}$$

Resolving vertical forces yields $F + R = W$ returning $F = 5W/8$, and taking moments about the base, $C - W(L/2) + RL = 0$, we have $C = WL/8$.

Recall that removing the support 'releases' the structure from its constraining redundancy and reveals the kinematical quantity of interest to compatibility. The redundancy must obviously be re-introduced for no load applied, in order to superpose to the original case: this is known as the *reactant* case for obvious reasons.

In Fig. 11.3(c) we choose instead C to be the redundancy. In releasing C, we permit free rotations of the base without vertical displacements. The original shear force at this end remains – preventing vertical movement, giving us equivalently a simple support: our released cantilever thus becomes a simply-supported beam.

The original loading produces a free clockwise rotation θ_1 at the base end equal to $WL^2/24EI$ from Table 10.1. We have chosen C to be anti-clockwise to negate the overall rotation, with θ_2 equal to $CL/3EI$ using our earlier result, giving $C = WL/8$ (as before) from $\theta_1 = \theta_2$.

The corresponding force reaction components are also shown in Fig. 11.3(c). With C, there is no net vertical force, hence equal and opposite components, C/L: the right-side component acts downwards to restrain the beam lifting off under the

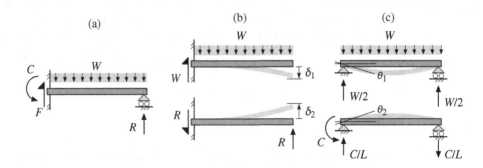

Figure 11.3 (a) Uniformly loaded propped cantilever. (b) Tip roller support in (a) removed to give a standard deflection case, with reaction force reinstated below without any loading: their separate tip deflections sum to zero. (c) Alternatively, rotation released at the built-in base to yield a simply-supported beam; the end moment is reinstated without loading, with equal and opposite rotations in both cases.

anti-clockwise tendency from C. This component opposes $W/2$ in the free case, giving R from (a) correctly equal in value to $W/2 - C/L$.

The choice of redundancy does not affect the final result, as noted, but it can make the analysis simpler. It does not matter here because the underlying free and reactant scenarios were all standard deflection cases, which we should generally aim for, as in the next example.

11.3   That End Moment (Again)

We re-draw the singly-redundant, two-span beam from Fig. 7.5(a) in Fig. 11.4(a). From the three general reactions, R_A, R_B and R_C, we might choose R_B to be redundant because by releasing it, we obviously have a simply-supported beam. However, as Fig. 11.4(b) shows, this case is not standard because W is not central, which demands a different solution approach.

A better approach splits the beam in two about the cental support to create two simply-supported beams. Rather than divide the central support in half, we excise free-bodies slightly bigger than one half of each side. Now the beams in Fig. 11.4(c) both have the reaction R_B, but just beyond it we must add to the ends unknown shear forces S_1 and S_2.

Considering two beams with different end points around the central support is not a problem provided local equilibrium is correctly postulated. If the beams are brought

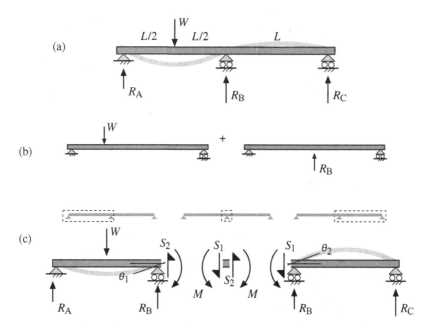

Figure 11.4 (a) Two-span beam. (b) Redundant central support reaction produces non-standard loading cases. (c) Alternatively, two separate halves produce standard loading cases.

together, they would overlap to create an elemental portion, where $R_B = S_1 - S_2$, see Fig. 11.4(c). This remains true when the width is condensed to zero and R_B conveys the step change in shear force, evident from Fig. 7.5(c) when R_B replaces λW.

The infinitesimal portions overhanging R_B must also bear equal and opposite bending moments, M, as shown, without, however, a jump in bending moment (because no moment is applied there).

Our choice of redundancy has to be linked to a common compatibility requirement. The exposed ends do not displace because of their adjacency to R_B, which rules out the shear forces (and R_B) as candidates, but they clearly share the same rotation. M is therefore our redundancy, and the response of each half can be described by standard cases.

For the left-side beam, the end rotation θ_1 depends on W and M, and nothing else since the forces R_B and S_2 act through the same point in the limit of zero element width. M, as we have drawn it, subtracts from θ_1, but acts in the same direction as θ_2 on the right-side beam. Table 10.1 again permits straightforward expressions:

$$\theta_1 = \frac{WL^2}{16EI} - \frac{ML}{3EI}, \quad \theta_2 = \frac{ML}{3EI} : \quad \theta_1 = \theta_2 \quad \rightarrow \quad M = \frac{3WL}{32}. \tag{11.2}$$

Taking moments about R_B for the left-side beam, we have $M + R_A L - WL/2 = 0$, giving $R_A = 13W/32$. Similarly, for the right beam, $R_C = -3W/32$, and resolving vertically for the original beam, $R_B = W - R_A - R_C = 22W/32$. The shear forces then follow from resolving vertically for each beam, $S_1 = 19W/32$ and $S_2 = -3W/32$.

If we construct the bending moment diagram, we have something similar to Fig. 7.5(c) after substituting for $\lambda W = R_B$. The gradients of this piece-wise linear profile quickly confirm the same values of S_1 and S_2 as well as the reactions R_A and R_C.

As with drawing earlier bending moment and shear force profiles, it is useful to sketch the displaced shape to prepare our expectations – in this case of the directions of θ_1 and θ_2. A formal sign convention for them is not necessary provided we import their directions correctly to the standard cases where, as we have seen, M and θ_1 were in opposite senses.

11.4   Frames (Again)

Monitoring compatible directions informally features in the next example, which does not immediately fit with any of the standard cases. A square portal frame in Fig. 11.5(a) is loaded centrally by a vertical force F. Given a uniform bending stiffness EI throughout, we wish to find the central deflection of the top beam and the reaction forces at the pinned feet.

Two simplifications stem from the symmetry of layout and loading, and hence response. Obviously, the vertical reaction forces are both $F/2$, with equal and opposite

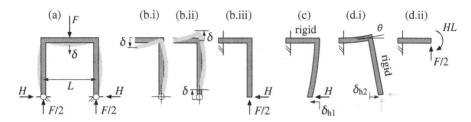

Figure 11.5 (a) Portal frame loaded centrally and responding symmetrically. (b.i) Deformation of half of the frame; (b.ii) rigid-body translation vertically of entire half-frame, to simulate clamping of the free end of top beam; (b.iii) equivalent structure with a free foot loaded by the original reaction forces. (c) First stage of deflection analysis. (d.i) Second stage, with the rotation depending on the tip forces in (d.ii).

horizontal reactions, H, which prevent splaying of the column legs. Second, because there is a plane of mirror symmetry about the line of action of F, only one half of the frame needs to be considered.

The boundary conditions for this sub-structure must also comply with the overall frame deflections. In the middle of the top beam, the gradient is zero but the vertical displacement, δ in Fig. 11.5(b.i), is not; each pinned foot is free to rotate but not to displace. An unknown bending moment acts over the middle 'free end' whose value, along with a known shear force of $F/2$, sets a middle gradient of zero. These stress resultants then determine δ, and overall equilibrium sets H.

But the question of how to use standard cases for this arrangement is still not obvious. When the deflected shape in (b.i) is translated upwards exactly by δ, the free end is neither rotated or displaced – as if fully built in. The vertical displacement can now be commuted to the pinned end, Fig. 11.5(b.ii), which can be treated as being free altogether under statically equivalent tip forces, H and $F/2$. Relative to the middle of the top beam, there can be no horizontal displacement, which becomes our compatibility equation for finding unknown H. The final version of the loaded half-frame is shown in Fig. 11.5(b.iii).

But there is one final embellishment for enabling analysis by standard cases. We assume the top beam to be rigid, and a first component, δ_{h1}, is due to the elastic deflection alone of the vertical column by a tip force, H, Fig. 11.5(c): this is the standard cantilever case rotated by $90°$ where $\delta_{h1} = HL^3/3EI$.

The second component, δ_{h2}, is due to the leveraged rotation of the now elastic top beam with the column effectively rigid, Fig. 11.5(d.i): $\delta_{h2} = \theta L$ from an anti-clockwise end rotation θ due to a clockwise moment HL and a shear force $F/2$, see Fig. 11.5(d.ii). The latter produces a component of rotation in the same sense as θ, returning $\theta = (F/2) \cdot (L/2)^2/2EI - (HL) \cdot (L/2)/EI$ overall.

We can now equate horizontal displacements, δ_{h1} and δ_{h2} to find $HL^3/3EI = FL^3/16EI - HL^3/2EI$ and hence $H = 3F/40$. Neglecting axial strains in the column and any shortening due to its small lateral deflections, the vertical displacement of each foot stems from the tip displacement of the cantilever portion alone, *i.e.*

$$\delta = \frac{(F/2)(L/2)^3}{3EI} - \frac{(HL)(L/2)^2}{2EI} = \frac{11FL^3}{960EI} \qquad (11.3)$$

after replacing H in terms of F: we have found our central deflection.

11.5   Final Remarks

The original structure is usually indeterminate. When sub-dividing into simpler elements, we seek standard cases loaded by the redundant quantities; these are applied on the cuts across the original structure when isolating free bodies, with their geometrical compatibility providing the extra conditions for a complete solution. This approach is typical of the general *Force Method* described by others. Assessing indeterminacy first of all is therefore necessary, and symmetry may help curtail the original structure into a simpler form.

12 Moment Distribution: At the Junction

We have noted before that raising indeterminacy by adding extra kinematical constraints promotes a higher structural stiffness. Their solution compared to their statically determinate counterparts also tends to be more involved because we must invoke compatibility as well as equilibrium (and their connection via the generalised Hooke's Law).

Our example later in this chapter contradicts the second statement because we constrain the original structure to behave more simply despite its new indeterminacy. There is also a curious 'inversion' of the equilibrium behaviour which then paves the way for thinking about how a familiar junction made from connected beam members behaves from a constraint point-of-view; in particular how bending moments are distributed around the junction.

12.1   The Determinate Case

When a couple (or moment), C, is applied to one end of a simply-supported uniform beam, its bending moment profile falls (or rises) linearly to zero at the other end from an initial step equal in height (or depth) to C. A constant gradient signifies constant shear force throughout; or equal and opposite (vertical) reaction forces, C/L, for a beam length L, see Fig. 12.1.

If the couple is now relocated along the beam, Fig. 12.2(a), the step in moment at C must also follow. No matter its position, however, the support reactions are unchanged from global equilibrium. The shear force profile is thus unaltered, as are the bending moment *gradients* either side of the step: as C is moved, the step appears to 'slide' along an equally inclined corridor, Fig. 12.2(b), whose parallel sides ensure a fixed step height.

Sub-dividing the beam into lengths L_1 and L_2 about C, Fig. 12.2(c), the salient bending moments C_1 and C_2 at the step can be verified as $(C/L) \cdot L_1$ and $(C/L) \cdot L_2$, respectively. Adding together, we obtain C again, which is precisely the equilibrium condition for an element of beam around C: equal gradients also stipulate $C_1/L_1 = C_2/L_2$.

We are also interested in how the overall beam stiffness varies with the position of C, which demands a solution of its deformation. The general deformed shape is easy to infer; to the left of C it curves downwards and *vice versa* for C anti-clockwise.

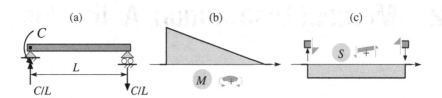

Figure 12.1 (a) Simply-supported beam loaded by end couple. (b) Bending moment profile. (c) Shear forces.

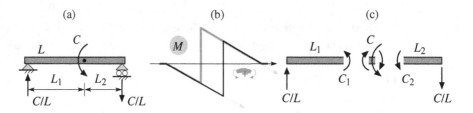

Figure 12.2 (a) Applied couple from Fig. 12.1 generally located along the beam. (b) Bending moment profile. (c) Equilibrium performance for an element around the couple.

Its displacements are less obvious, in particular, the vertical movement *at* the point of application of C.

There is clearly none when C acts at either support; and none when C is central from anti-symmetrical considerations, see Fig. 12.3(a) and *c.f.* Chapter 9. Each rotation at C is a standard result from Chapter 11: $CL/3EI$ for C at either support; for the middle case, we have equivalently $C/2$ applied to the ends of simply-supported beams of length $L/2$, Fig. 12.3(b), giving a fourfold reduction in rotation of $(C/2) \cdot (L/2)/3EI$.

The middle case is clearly four times *stiffer* in rotation compared to the end-loaded case, both of which define the stiffness extrema for the case of C anywhere else. None of the standard cases apply nor the method of Deflection Coefficients (*c.f.* Chapter 11) in a simple way because the vertical deflection at C is another unknown quantity. Consider instead the governing differential equation approach of Chapter 10.

First, we specify a general span-wise coordinate x from the left-hand end up to but not beyond the position of C, see Fig. 12.3(c). For positive moments hogging, $M(x)$ is $-(C/L)x$ and, for a positive upwards transverse displacement v_1, $M(x) = -EId^2v_1/dx^2$. Integrating yields the rotation and then the displacement expressions:

$$\frac{dv_1}{dx} = \frac{Cx^2}{2EIL} + a_1 \quad \text{(a)}, \quad v_1 = \frac{Cx^3}{6EIL} + a_1x + a_2 \quad \text{(b)}. \quad (12.1)$$

The coefficient a_2 is zero given that $v_1 = 0$ at the left end: to find a_1, we need to match the deformation at C expressed by the coordinate y moving to C from the right-hand end, Fig. 12.3(c).

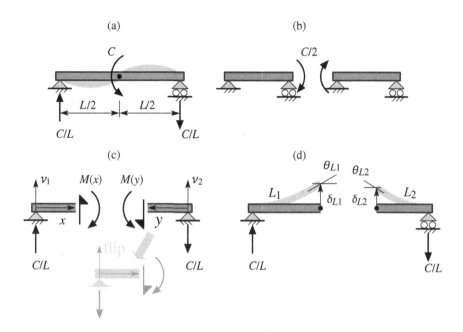

Figure 12.3 (a) Couple applied centrally to a simply-supported beam. (b) Statically equivalent half-beams. (c) General beam free-body sections for displacement analysis. (d) Corresponding compatibility conditions at the point of application of the couple in (a).

The transverse displacement is now v_2 and the exposed bending moment, $M(y)$. The governing equation of deformation is similarly written in terms of y and integrated, but recognise that if we 'flip' this beam portion in Fig. 12.3(c) to the back-side view, we have the same description as the x portion with an opposite-sense reaction, *i.e.*

$$\frac{dv_2}{dx} = -\frac{Cy^2}{2EIL} + b_1 \quad \text{(a)}, \quad v_2 = -\frac{Cy^3}{6EIL} + b_1 x \quad \text{(b)}. \qquad (12.2)$$

These coordinates meet at C where the displacements from Fig. 12.3(d), $\delta_{L1} = v_1(L_1)$ and $\delta_{L2} = v_2(L_2)$, are equal. Their corresponding rotations, θ_{L1} and θ_{L2}, are also equal but of opposite signs for continuity in the deformed gradient of the beam. Substituting for $x = L_1$ and $y = L_2$ into the expressions above, we find:

$$\frac{CL_1^2}{2EIL} + a_1 = \frac{CL_2^2}{2EIL} - b_1; \quad \frac{CL_1^3}{6EIL} + a_1 L_1 = -\frac{CL_2^3}{6EIL} + b_1 L_2. \qquad (12.3)$$

The coefficients, a_1 and b_1, can be solved between these two equations, for example, $a_1 = (C/6EIL^2) \cdot (L_1^3 + 3L_1^2 L_2 - 2L_2^3)$. Substituting back into Eqs. (12.1(a) and (b)) and using the compact definitions, $L_1/L = \lambda$ and $L_2/L = 1 - \lambda$ from $L_1 + L_2 = L$, it can be shown that:

$$\theta_{L1} = -\theta_{L2} = \frac{CL}{3EI} \cdot \left[\lambda^3 + (1-\lambda)^3\right] \text{ (a)},$$

$$\delta_{L1} = \delta_{L2} = \frac{CL^2}{3EI} \cdot \left[\lambda(1-\lambda)(1-2\lambda)\right] \text{ (b)}. \tag{12.4}$$

λ lies in the range zero to unity when C is moved from one end to the other, and we confirm a fourfold reduction in rotation and zero displacement in the middle where $\lambda = 1/2$. In general, each rotation is symmetrical about the centre in keeping with the fixed direction of C.

Denoting θ_{L1} equal to θ in the same sense as C, the rotational stiffness from C is the ratio of C to θ. A convenient dimensionless form is $\bar{C} = (C/\theta) \cdot (L/3EI)$, now equal to $1/[\lambda^3 + (1-\lambda)^3]$. When plotted later in Fig. 12.5(a), $\bar{C}$ rises symmetrically from unity at either end to a value of four in the middle.

The displacement varies anti-symmetrically, with C moving upwards in the first half and downwards, second. The same maximum value in either half occurs at positions given by $\lambda = 1/2 \pm 1/\sqrt{12}$ (from differentiating Eq. (12.4) and setting equal to zero).

12.2   The Indeterminate Case

The previous displacement at C is now prevented by connecting the original unloaded beam to a new roller support located at C, Fig. 12.4(a), before the couple is applied again. The couple divides into its usual pair, C_1 and C_2, applied to each portion, which behave as shorter end-loaded, simply-supported beams. The end rotations either side of the internal support are, thus, $\theta_1 = C_1L_1/3EI$ and $\theta_2 = C_2L_2/3EI$.

The rotations are equal to each other for the drawn directions, which sets $C_1L_1 = C_2L_2$, or $C_1/L_2 = C_2/L_1$, and the bending moment profile in Fig. 12.4(b): compare

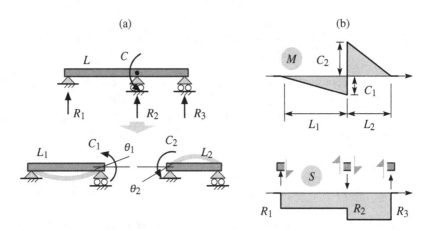

(a)   (b)

Figure 12.4 (a) Extra roller support added to the position of the applied couple in Fig. 12.3. The indeterminate beam can be directly divided into two smaller, end-loaded beams. (b) Corresponding bending moment and shear force profiles.

this to our original simply-supported case where $C_1 L_2 = C_2 L_1$. Here, the gradients either side of C *viz.* shear forces are generally different, giving a non-zero internal reaction force, manifest as the step in Fig. 12.4(b).

We can write C_2 equal to $C_1(L_1/L_2)$, in order to find C_1 explicitly from $C = C_1 + C_2$; knowing that $L = L_1 + L_2$, then $C_1 = CL_2/L$. Enunciating our equal rotations as θ again, we can show $\theta = CL_1 L_2/3EIL$, or in the same dimensionless form: $\bar{C} = (C/\theta) \cdot (L/3EI) = 1/\lambda(1 - \lambda)$.

This variation is plotted in Fig. 12.5(b) alongside the originally determinate variation (in grey). The stiffness everywhere has increased except for the middle where the curves coincide. At each end, the stiffness rises asymptotically, signifying zero rotations in the limit when C acts at either end.

This seems to be an odd result given that the internal roller support has effectively merged with that at the end to give a determinate beam. But imagine what happens as C approaches, say, the left end in Fig. 12.4 very closely and the internal support remains in place. L_1 is much smaller than L_2, and C_1 is approximately C from $C_1 = CL_2/L$.

Also recall that θ_1, our rotation at C, is proportional to C_1 *and* to L_1. Despite most of C being applied to the left side beam, its shortness sets a low rotation that ultimately tends to zero. The way in which C divides in the determinate case is reversed, meaning that most of C is applied to a larger unconstrained portion of length L_2.

By adding a roller support, we have increased the rotational beam stiffness through indeterminacy. Crucially, we have seen that compatibility alters the way in which the applied moment is carried on either side: its *distribution* in view of equilibrating bending moments is now set by $C_1 L_1 = C_2 L_2$ compared to $C_1 L_2 = C_2 L_1$ from the determinate case.

The former result is mirrored by the relationship between the reaction forces R_1 and R_2 in a simply-supported beam loaded somewhere along by a point force. Taking moments underneath the load, we find $R_1 L_1 = R_2 L_2$: the same is also 'infinitely' stiff if the force is applied over a support, and minimally so in the middle.

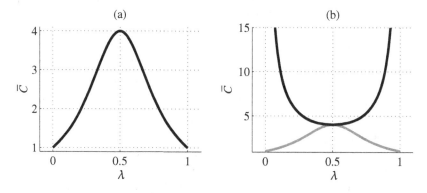

Figure 12.5 (a) Dimensionless rotational stiffness, $\bar{C}$, of the determinate beam from Fig. 12.2: λ is moment position from the left-hand end. (b) The same for the indeterminate beam from Fig. 12.3, black, compared to (a), grey.

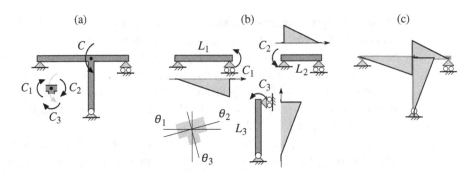

Figure 12.6 (a) T-frame loaded by couple at the beam junction. (b) Statically equivalent beams, which are simply supported and end loaded to the same rotation. (c) Bending moment profile throughout.

Even though the indeterminate case is split into two simply-supported beams, we can think of their merging as forming a moment-carrying *junction* over a single internal support. We do not have to limit ourselves, therefore, to two horizontal beams provided each beam is supported simply in the sense of their own transverse displacements.

An example of a three-beam junction appears in the T-frame in Fig. 12.6(a). For convenience, again, a couple C is applied to the junction whose elemental equilibrium suggests that C equals the sum of the immediate bending moments, C_1, C_2 and C_3.

The left-hand end is simply supported and the right side is a roller support, as per the ends in Fig. 12.4. The junction itself must behave as if it were simply supported so that the horizontal portions on either side can be treated equivalently. Fundamentally, there can be no vertical displacement, which is accorded by the column being pin-supported at its base *and* by neglecting any axial extensions of the column. The latter arises because the column now applies a 'support' force to the horizontal beam, giving a column force by reaction.

By the same measure, we can neglect horizontal extensions of the top part so that the column is effectively simply supported at its ends. The arrangement can now be separated into three beams, Fig. 12.6(b), loaded by end moments C_1, C_2 and C_3, as shown.

Each end rotation is now a standard result with $\theta_1 = C_1L_1/3EI$, $\theta_2 = C_2L_2/3EI$ and $\theta_3 = C_3L_3/3EI$ when all parts have the same EI. These are equal, which sets $C_1L_1 = C_2L_2 = C_3L_3$ and thence the distribution of bending moments given the overall geometry; their general linear profiles are plotted in Fig. 12.6(c) on the tensile side of each part.

As a final example, consider when $L_1 = L_2 = L/2$, and when L_3 is some proportion γ of L: we wish to find the stiffness of a symmetrical T-frame as a function of its column length. Equal rotations admit $C_1/2 = C_2/2 = C_3\gamma$ and $C = C_1 + C_2 + C_3 = C_1(1 + 1 + 1/2\gamma)$, whence $C_1 = 2C\gamma/(4\gamma + 1)$. We have $\theta_1 = C_1L_1/3EI$ which, after substitution and rearrangement, returns $\bar{C} = (C/\theta_1) \cdot (L/3EI) = 4 + 1/\gamma$.

When γ is very large, the column is long and $\bar{C} \approx 4$. This is the same value as the roller-supported beam in Fig. 12.4 when $\lambda = 1/2$: the rotational stiffness of the column itself is negligible in comparison. As γ decreases, $\bar{C}$ increases rapidly under the reciprocal nature. Indeed, for $\gamma = 1/4$, the stiffness is doubled compared to beam case alone: the addition of a quarter-length beam as a column, with its supports, has a disproportionate and perhaps favourable outcome for the joint performance if a higher stiffness is preferred.

12.3   Final Remarks

We have deliberately chosen a loading couple at the joint in all examples, for this simplifies equilibrium and permits immediate insight using standard results. Other loads will introduce other compatibility conditions alongside those due to the joint bending moments, but the analysis remains the same. If no couple is applied to the joint, we stipulate $C_1 + C_2 + C_3 \ldots = 0$ and work from there.

Our analysis has also been linear elastic, and geometrical compatibility has been key. Later, in Chapter 16, where we deal with Lower Bound design (again), we can relax the compatibility stringency, but the sense of moment distribution remains key. We have also noted how the column carries an axial force: if compressive, axial buckling at a smaller force than needed for permanent material yielding may become an issue.

Part V

Design Choices

13   Bending *vs* Stretching

Our view of structures has been greatly simplified so that we can write accurate but compact expressions for their behaviour. Furthermore, we have, thus far, only dealt with planar beam structures where *torsion* – arising indirectly from out-of-plane loads, is largely remiss: we shall return to this later in Chapters 14 and 17.

Central to our view is member slenderness whose length is greater than its depth, with wall thicknesses even smaller if members have thin-walled cross-sections. Their constitutive structural response is accurately captured by a one-dimensional view of elastic displacements; that bars compress or extend axially whereas beams and columns displace transversely.

13.1   A Question

Do these modes ever conflate in the sense of overall displacements? To answer matters, consider the simplest pin-jointed square arch in Fig. 13.1(a), which carries a vertical point force. It must displace symmetrically, and the exaggerated deformed shape shows linearity, *i.e.* each bar extending along its own length and also rotating.

The arched layout of structure and geometry is preserved in Fig. 13.1(b) after replacing the bars with built-in beams; the loaded joint, however, remains pinned. Clearly, its displacement stems from vertical components of axial displacements combined with those from transverse curving of the beam.

The built-in arch is statically indeterminate, as described shortly. Base rotations are curtailed; yet, if the bars are slender enough, it is customary to treat the arrangement as pin-jointed throughout and, thence, statically determinate. What is the validation for doing so? More specifically, we may ask about the relative balance of transverse displacements arising from bending of the beam against those from a bar rotating in (a). If this is very small, bending effects can be neglected, and the modes do not conflate.

We solve matters momentarily, but beforehand we note a more subtle coupling of bending and extension in Fig. 13.1(c). The arch is pinned in the usual places but is now initially curved. When the central joint moves upwards, the straight lines connecting it to the base joints can freely rotate about them. The extension between them, however, is achieved by each half becoming less curved and thus bending in the opposite sense of its initial curvature. There is, of course, some direct extension along the curved

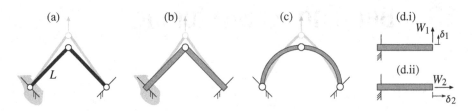

Figure 13.1 (a) Pin-jointed square arch. (b) Inclined cantilever arch with pinned loading joint. (c) Curved beam arch. (d.i) Transversely tip-loaded cantilever; (d.ii) axially loaded cantilever.

centre-lines, but this will be insignificant for slender bars compared to the separation induced by 'unbending.'

13.2   A Formal Solution

Bending and extension operate separately from each other in Fig. 13.1(b) because each half can only carry orthogonal components of tip forces, as we shall see. It is equivalent to the uniform cantilever in Fig. 13.1(d) loaded transversely by a tip force, W_1, then axially by another, W_2; for small displacements, W_2 produces no bending and W_1 no extension.

Their tip displacements δ_1 and δ_2 are standard expressions, $\delta_1 = W_1 L^3 / 3EI$ and $\delta_2 = W_2 L / AE$. As usual, E is the Young's Modulus, A is the cross-sectional area and I is the second moment of area, where I can be defined in terms of a radius of gyration for the cross-section, k, as $I = A k^2$. It is convenient to define a parameter $\beta = (1/3) \cdot (L/k)^2$ where L/k is a measure of the beam slenderness given that k is some proportion of the cross-sectional depth; the $1/3$ pre-factor adds further convenience shortly.

Evidently $\delta_1 / \delta_2 = \beta$ when $W_1 = W_2$, and transverse deflections can be much larger than axial extensions for the same size of loading, by a factor of 30 and upwards when $L/k \approx 10$ and larger. Conversely, $1/\beta$ expresses the ratio of transverse-to-axial tip forces required for equal displacements. Recognise also the more general expression, $(W_2/\delta_2)/(W_1/\delta_1)$, the ratio of axial-to-transverse stiffness, which is also equal to β.

Solving for the indeterminate middle example, we use equilibrium and compatibility as per normal: forces and displacements both matter, as does their relative contribution, imbued in β. First, let us calculate the true pin-jointed response of the arch in Fig. 13.1(a) before returning to Fig. 13.1(b).

Free-body equilibrium of the loaded joint gives equal bar tensions of $P/\sqrt{2}$. The bar extensions e are also the same and equal to $(P/\sqrt{2})L/AE$. The loaded joint displaces vertically with orthogonal components due to extension and rotation of each bar, which are drawn as the vector diagram in Fig. 13.2(a) with respect to the left bar. The original inclination of either bar at $45°$ sets $\delta = \sqrt{2}e$, which is equal to PL/AE after replacing e.

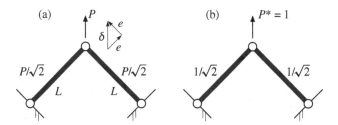

Figure 13.2 (a) Response of pin-jointed arch. Bar tensions are $P/\sqrt{2}$ and the inset shows a vector diagram of displacement components normal and parallel to each bar: all inclinations are $\pm 45°$. (b) Virtual equilibrium set.

We can also confirm using the method of Virtual Work from Chapter 5, where the actual kinematics are of interest. We therefore apply a virtual force of unity to the loaded joint in Fig. 13.2(b), resulting in $T^* = 1/\sqrt{2}$ for each virtual bar tension. Equation (5.1) can now be written $1 \cdot \delta = 2T^* e = (2/\sqrt{2}) \cdot (PL/\sqrt{2})/AE$, whence $\delta = PL/AE$ again.

The built-in arch is re-drawn in Fig. 13.3(a). Splitting in half produces two statically determinate cantilevers now tied by unknown axial and transverse forces, T and S, at each tip; no moment can be transmitted across the pin-joint.

Vertical equilibrium of this joint in Fig. 13.3(b) gives $P = \sqrt{2}T + \sqrt{2}S$, with respective displacement components d_T and d_S equal to TL/AE and $SL^3/3EI$. These are orthogonal and inclined at $45°$ to the horizontal, and sum symmetrically for a vertical displacement, δ, setting $d_T = d_S$ and $\delta = d_T/\sqrt{2} + d_S/\sqrt{2}$.

In order to compare the final displacement to that of the pin-jointed arch, we focus on T and d_T. Since $d_T = d_S$, we can write $TL/AE = SL^3/3EI$, which turns out to have the compact form $T = \beta S$. In our original equilibrium statement, P now equals $\sqrt{2}T(1 + 1/\beta)$ after substituting for S; and $\delta = \sqrt{2}d_T$ after replacing d_S with d_T, giving $\delta = \sqrt{2}TL/AE$.

Eliminating T between the expressions for P and δ:

$$T = \frac{P}{\sqrt{2}\left(1 + 1/\beta\right)} = \frac{\delta}{\sqrt{2}} \frac{AE}{L} \quad \rightarrow \quad \delta = \frac{PL}{AE} \cdot \frac{\beta}{1 + \beta}. \tag{13.1}$$

Our pin-jointed result is now multiplied by the factor $\beta/(1+\beta)$. This factor is smaller than unity for $\beta > 1$, giving a smaller displacement because bending also resists P. When β takes a typical value of 30 or more ($L/k \geq 10$), the factor is, however, almost equal to unity, suggesting a very small departure from our pin-jointed result outright.

The inclusion of bending, therefore, makes little difference to the final result. Bending displacements are equally viable compared to extensions, and equilibrium apportions equal levels of components, $\sqrt{2}S$ and $\sqrt{2}T$, but it is the stiffness, or rather the lack of it, that is responsible. An increasingly slender beam offers less transverse bending stiffness compared to axial, and we can justifiably replace each built-in base – the source of bending – by a pin-joint, in order to arrive back at our original truss.

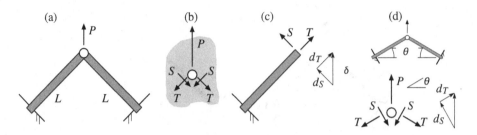

Figure 13.3 (a) Loaded built-in square arch. (b) Equilibrium of loaded pin-joint. (c) Same forces applied to the tip of each cantilever leg, and their respective displacement components. (d) Built-in arch of general initial inclination, and associated directions of tip forces and displacement components.

13.3   Validity Range

But how does our conclusion hold for other geometries of initial layout, which alter the equanimity of both force and displacement components during solution? Consider thus when the built-in arch, Fig. 13.3(d), is now generally inclined at angle θ to the horizontal but retains the same bar lengths and individual stiffness properties as before.

Vertical equilibrium is straightforward and compatibility yields a net vertical displacement, δ, and none horizontally, to present:

$$P = 2T \sin \theta + 2S \cos \theta \quad \text{(a)}; \quad \delta = d_S \cos \theta + d_T \sin \theta \quad \text{(b.i)},$$

$$d_S \sin \theta = d_T \cos \theta \quad \text{(b.ii)}. \tag{13.2}$$

The individual axial and transverse tip displacements, d_T and d_S, have the same expressions in T and S, and Eq. (13.2(b.ii)) can be re-written as $T \cos \theta = \beta S \sin \theta$ after substituting them along with β; or that $T = \beta S \tan \theta$. Compared to the square arch, T is modified by $\tan \theta$, being amplified or reduced either side of $\theta = 45°$.

Either of S or T can be replaced in Eq. (13.2(a)) with its corresponding component in Eq. (13.2(b.i)) and, between them, solved explicitly for W and δ. For example:

$$P = 2T \sin \theta + 2 \left(\frac{T \cos \theta}{\beta \sin \theta} \right) \cos \theta; \quad \delta = \left(\frac{d_T \cos \theta}{\sin \theta} \right) \cos \theta + d_T \sin \theta. \tag{13.3}$$

Writing the first expression explicitly in terms of T and the second in d_T, and using $d_T = TL/AE$, we arrive at the relationship between δ and P as a function of θ:

$$\delta = \frac{PL}{AE} \frac{\beta (\sin \theta + \cos \theta \cot \theta)}{2 (\beta \sin \theta + \cos \theta \cot \theta)} \quad \rightarrow \quad \delta = \frac{PL}{AE} \cdot F_T. \tag{13.4}$$

The left-hand side has been re-written on the right using a dimensionless factor, F_T, in θ. It multiplies our earlier pin-jointed result for the square arch (assuming A, E, L and P are unchanged), and it is also the reciprocal of overall (dimensionless) stiffness, or *compliance*, of the arch written explicitly in terms of the extensional performance: δ in Eq. (13.1) is also confirmed for $\theta = 45°$.

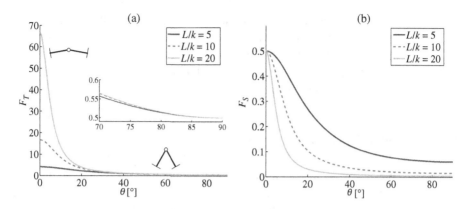

Figure 13.4 Variation of compliance factors in extension (F_T), (a), and in bending (F_S), (b), with arch inclination θ, for different slenderness ratios, L/k.

The complementary view in bending considers S and d_S instead when finding δ and P, with a compliance factor, F_S:

$$\delta = \frac{PL^3}{3EI} \frac{\cos\theta + \sin\theta\tan\theta}{2(\cos\theta + \beta\sin\theta\tan\theta)} = \frac{PL^3}{3EI} \cdot F_S. \tag{13.5}$$

Recognise that the pre-factor is the standard deflection for a tip-loaded cantilever. If we consider plotting either F_S or F_T by themselves – as functions of θ, we garner each structural performance in relative terms as the arch inclination changes. Both are, therefore, plotted as a series of curves for typical slenderness values ($L/k = 5, 10, 20$) from flat ($0°$) to tall ($90°$) arches in Figs. 13.4(a) and (b).

Consider first when $\theta = 0°$. Each initial value of F_T in Fig. 13.4(a) is different, but are all well above unity and equal to $\beta/2$ from directly calculating. The extensional compliance for this layout is therefore low, giving a displacement considerably higher than the pin-jointed square arch.

Furthermore, because the layout is flat, there can be no axial tension in response to P, and each cantilever half carries $P/2$ as a transverse tip force. The displacement now is simply calculated, $\delta = (P/2)L^3/3EI$, which accounts for all curves emerging from the same point, $F_S = 0.5$, in Fig. 13.4(b), *c.f.* Eq. (13.5).

As θ increases, the capability of carrying more axial force builds up. F_T rapidly decreases, as does the displacement, with a more pronounced rate of decrease for larger values of slenderness. We expect this from what we know about the square arch and Eq. (13.1) where a larger β, hence slenderness, reduces the bending displacement component in relative terms.

All curves then converge around $\theta = 30°$, becoming asymptotic to $F_T = 0.5$, Fig. 13.4(a). Again, the bending contribution is now diminishing across all curves because δ is being correlated to extensional performance: axial effects, we say, now *dominate*. Bending effects do not disappear altogether, resulting in very small differences between the curves in the inset plot.

Figure 13.4(b) also confirms the same: reading vertically, the curves are reversed in order compared to (a), which is expected given that stockier beams have a relatively higher bending stiffness; but even the stockiest beam curve ($L/k = 5$) falls away quickly. At $\theta = 90°$, there is no bending because S is zero, but there is still some extension and, thus, a non-zero δ (and F_S).

13.4   Final Remarks

The initial geometry clearly dictates the dominant response. A flatter arch prefers to bend, but even modest inclinations are led by extensional behaviour. This is promoted by initially straight beams and a pin-jointed loading, which minimises bending moments throughout.

In most other problems, the initial geometry and loading profile dictate the dominant responsive mode unambiguously. Quite simply, if there are transverse forces applied anywhere other than a joint, the stiffness (or compliance) of the structure will depend on the local displacements they produce away from any joints, pinned or otherwise, and in this sense, bending is pre-eminent for slender members.

14 Cross-Sectional Stiffness

Relatively small deflections allow us to write accurate equilibrium statements using the initial, unloaded geometry. But how 'small' is small? Automatically, our structure is stiff enough so that vibrations, say from footfall or wind loading, do not accelerate its fatigue life: that excessive strains verging on plastic limits are avoided; or that we simply do not notice movements for our own sense of safety.

Our question is, therefore, linked to stiffness. Thinking of how discrete structures are built from beams and bars, there is the fundamental material stiffness given by the Young's Modulus E (and the shear modulus, G, and Poisson's Ratio, ν). There is, then, the first structural incarnation of cross-sectional stiffness defined by each member's constitutive response mode and finally, the overall, aggregated structural stiffness.

There is a clear hierarchy of stiffness, and earlier Chapters 5, 10 and 11 focussed on calculating the load-deflection relationships in bespoke trusses and frames. Material properties are invariant, or *structurally insensitive*, leaving cross-sectional stiffness in a pivotal role.

The axial stiffness of a pin-jointed bar of length L and area A of cross-section is the familiar EA/L from the ratio of applied force to extension (or compression). To maximise stiffness in accordance with small bar extensions and nodal displacements, we can increase A (aside from E) and reduce L, but these make the bar heavier *or* lighter.

Incorporating the mass of the element in some sort of *specific stiffness* measurement, therefore, has merit; for example, we might want to extract as much stiffness from as little material as possible by maximising their ratio – see later.

But because a bar generates, essentially, uniform stresses everywhere (in our assumption), there is little else we can do at the level of cross-section, for its shape is largely irrelevant. The story is, however, different with beams, which generate varying cross-sectional stresses from bending and shearing. Keeping the mass of the structure down is accomplished preemptively using thin-walled, hollowed sections, which in turn introduces extra, often simplifying, assumptions about how local stresses vary within its thin walls.

The cross-sectional level parameters are the well-known stiffness terms, EI in bending and GJ in torsion. Computing the second moment of area, I, and the polar moment of area, J, are therefore central to our stiffness assessment, and we start with informal calculations, for saving time as well as reinforcing associated principles.

We then return to our original question about the level of stiffness required for ensuring small deflections before a final, brief discussion on why transverse displacements are determined mainly from curving and not vertical shearing of the beam locally.

14.1   Second Moment of Area

The rectangular hollow cross-section in Fig. 14.1(a) has outer dimensions of $2a \times 2b$ aligned to principal axes x and y. It has a uniform wall thickness t, setting the inner peripheral size to be $2(a - t) \times 2(b - t)$.

This is the second simplest type of bending cross-section; the first is a solid rectangular section, where the second moment of area about the x-axis, denoted as I_{xx}, is the familiar '$bd^3/12$', b being the breadth and d the depth.

The hollow cross-section can be treated as two 'solid' rectangles: of an inner core *subtracted* from the outer shape. Calculating the overall, I_{xx} is the same straightforward subtraction because both centroids of shape are collocated on the x-axis, otherwise the *Parallel Axis Theorem* has to be deployed. Thus, we find:

$$I_{xx} = \frac{2a \cdot (2b)^3}{12} - \frac{2(a - t) \cdot (2(b - t))^3}{12} = \frac{4}{3}\left[ab^3 - (a - t)(b - t)^3\right]. \quad (14.1)$$

In a typical thin-walled section, the thickness is small compared to the overall size, enabling us to take t/a and t/b to be much less than unity. Second-order terms leading in t^2 can be neglected, and using the first two terms of the Binomial Theorem, $(b - t)^3$ is approximately $b^3(1 - 3t/b)$. As a result, $I_{xx} \approx (4/3) \cdot (b^3t + 3ab^2t)$.

These two terms have a clear physical meaning after recalling the Parallel Axis Theorem, Fig. 14.1(b). For an area, A, of part of the cross-section located q normal to the axis concerned, its I_{xx} portion comprises $A q^2 + I_G$, the latter being I about its own centroid in the same direction.

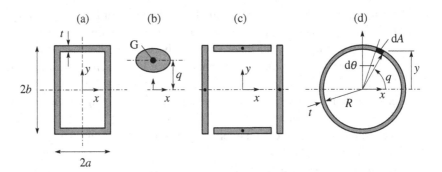

Figure 14.1 (a) Thin-walled rectangular hollow section. (b) Element of area for parallel axis calculation about x-axis. (c) Constituent plate elements for (a) with their centroids as dots. (d) Thin-walled circular hollow section.

We can split our thin-walled section into its four side 'plates' in pairs of length $2a$ and $2b$ without changing their vertical positions, Fig. 14.1(c). For the top and bottom plates, I_G is $2at^3/12$ and $A\,q^2$ is $2at \cdot b^2$. The difference between them is thus $(t/b)^2$ and I_G is negligible in comparison. Each vertical plate is already central with only I_G non-zero and equal to $t(2b)^3/12$. Adding all plate contributions we arrive at I_{xx} equal to $2 \times 2ab^2t + 2 \times 8tb^3/12$, as earlier.

A square thin-walled section with $a = b$ therefore has I_{xx} equal to $16b^3t/3$, *c.f.* its 'solid' counterpart of $4b^4/3$. By making the section thin, its second moment of area is reduced by a factor of t/b, which can be very small. But its *material* area and hence mass (per unit length of section) reduces by the same factor, giving comparable specific stiffness values. The particular dimensional dependencies on b and t for both also help answer a question about the circular tube shown in Fig. 14.1(d).

Its I_{xx} is calculated formally from $\int y^2 \mathrm{d}A$. The ordinate y is $R \cos\theta$, where R is the radius of the tube to mid-thickness and θ is an intrinsic coordinate measuring the position of an elemental area, $\mathrm{d}A$, from the x-axis. Clearly, $\mathrm{d}A = R\mathrm{d}\theta$, and the integrand is $R^3 \cos\theta\mathrm{d}\theta$ between limits of zero and 2π, returning $I_{xx} = \pi R^3 t$.

This contains the same product of 'size' parameter, *i.e.* R to the power of three and thickness as the square hollow section. The second moment of area for a solid circular tube must vary with the fourth power of R in comparison, and if we subtract, again, a solid core for calculating the thin-walled result, each contribution to I_{xx} depends on R^4 and $(R-t)^4$. Denoting the proportion of variation as β, then $I_{xx} = \beta R^4 - \beta(R-t)^4$, which must be equal to $\pi R^3 t$. Taking t/R to be much less than unity, β works out at $\pi/4$, giving I_{xx} for a solid circular tube of radius R as $\pi R^4/4$ without any further calculation.

Returning to the thin-walled result, we noted that R was measured to the 'mid'-thickness. But often with very thin sections, this is not convenient because of the change in scale – usually – in units (*e.g.* a radius in cm *vs* a thickness in mm). If, instead, we measure R to the outer radius, I_{xx} is equal to $\pi R^4/4 - \pi(R-t)^4/4$; if inner, then $I_{xx} = \pi(R+t)^4/4 - \pi R^4/4$. Multiplying out both expressions and simplifying as before, we arrive at the same I_{xx}. The exact end-point of the radius does not matter provided it lies within the wall thickness.

Viewing I_{xx} as a scalar quantity, which can be subtracted for hollow sections of standard shapes, provides a valuable analytical short-cut. But the formal definition of I_{xx} disregards how the area of shape is distributed in the x direction; only the y-wise distribution matters. Consequently, we may be able to artificially 'distort' the x-wise distribution of shape to our advantage.

For example, the narrow trapezoidal and solid cross-section in Fig. 14.2(a.i) is inclined at angle θ to the horizontal x-axis about its centroid. It has length L and through-thickness t, and an element of area, $\mathrm{d}A$, is highlighted at height y normal to x.

$\mathrm{d}A$ is also trapezoidal but it can be excised and sliced vertically into two pieces, then rejoined on their slanting faces, to produce a rectangular element, Fig. 14.1(a.i). There is no change in its area and none in y: we can repeat this process for every elemental area above and below. The entire arrangement can then be shunted left or right against the same vertical line without changing the y coordinate of each element.

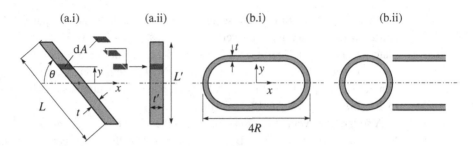

Figure 14.2 (a.i) Inclined, thin strip ($t \ll L$); (a.ii) equivalent rectangular strip for calculating I_{xx}. (b.i) Squashed thin-walled cylinder; (b.ii) equivalent simpler cross-sections for I_{xx}.

There is now a rectangular solid section of depth L' and width t', equal to $L \sin \theta$ and $t / \sin \theta$, respectively: I_{xx} is the standard result, $t' L'^3 / 12$, equal to $(t L^3 / 12) \sin^2 \theta$. This result is well-known for a rotated rectangular rod, which differs from the present by having right-angled ends. But their extent is negligible when t is much less than L, indeed when θ is zero, I_{xx} is zero because the strip is now a thin line along x.

The same logic can be applied to the squashed cylindrical cross-section in Fig. 14.2(b.i), made from two semi-circular halves joined by horizontal strips of width $2R$. Everywhere the through-thickness is t, and I_{xx} can consider the circular ends as a single cylinder separated from the strips, Fig. 14.2(b.ii).

The former is our standard result, $\pi R^3 t$, and neglecting I_G of the strips, they contribute $2 \times (2Rt)R^2$ via the Parallel Axis Theorem: overall $I_{xx} = (\pi + 4)R^3 t$. Note that the total length of the strips is $4R$, compared with $2\pi R$ for the ends: despite being shorter, they account for more than half of I_{xx} because their areas are located further away from the x-axis on average compared to the semi-circles.

14.2   Bi-axial Bending

Multiplying I_{xx} by E produces our cross-sectional bending stiffness when a moment is carried along the x-axis within a symmetrical cross-section. If there is a second orthogonal component in the y direction, $E I_{yy}$ is its respective stiffness, and the combined response depends on both the cross-sectional geometry and the weighting between the moment components, which now constitute *bi-axial bending*.

The moments, however, can always be combined into a single resultant not parallel to either axis; recall that we can treat all moments as double-headed vectors operating positively under Maxwell's right-hand screw rule, where the thumb points along the axis of bending and fingers curve in the sense of bending. If the resultant magnitude is M and inclination to x is θ, then $M = \sqrt{M_x^2 + M_y^2}$ and $\tan \theta = M_y / M_x$ for original x- and y components, M_x and M_y, Fig. 14.3(a).

A resultant brings us back to single-axis bending and the possibility of a 'composite' EI combining I_{xx} and I_{yy} in some meaningful way, with I as the equivalent

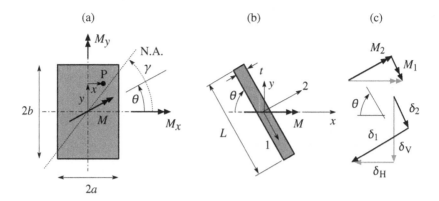

Figure 14.3 (a) Solid rectangular cross-section carrying a bending moment M inclined at angle θ to the x-axis: the neutral axis is inclined at γ. (b) Inclined rectangular strip carrying a horizontal moment. (c) Components of bending moment and displacement parallel (1) and normal (2) to the strip axis, with corresponding vertical and horizontal displacements.

second moment of area. For this approach, we must understand the properties of the current neutral axis – our usual datum for any I.

We can superpose the axial stresses generated separately by M_x and M_y, which must consider a general point P located at (x, y) within the cross-section, which we take to be a solid rectangle, $2a \times 2b$, in Fig. 14.3(a) for simplicity. Above the x-axis, M_x generates tensile stresses and *vice versa*, and to the left of y, M_y produces the same, hence the tensile stress at P is $M_x y / I_{xx} - M_y x / I_{yy}$.

If P lies on the neutral axis there is no stress, setting y/x equal to $(M_y/M_x) \cdot (I_{xx}/I_{yy})$. The first ratio is $\tan \theta$ and the second is $(b/a)^2$ knowing $I_{xx} = 4ab^3/3$ etc. The neutral axis is another line different from the x- and y axes but also passing through the centroid; if its inclination to x is shown as γ in Fig. 14.3(a), then $y/x = \tan \gamma$; or $\tan \gamma = (b/a)^2 \tan \theta$.

Only when b and a are equal do the neutral and bending axes coincide: there is no simple correlation between them and thus no obvious way of viewing a composite bending stiffness apart from this case or when the rectangle is very narrow with, say, $b \gg a$. The neutral axis now approaches the y-axis because $\tan \gamma$ is very large and, no matter the proportion of M_x and M_y, the cross-section stiffness will defer to the much weaker I_{yy} with a higher displacement in the x direction compared to vertical.

We can view this another way using a single horizontal moment, M, applied to a thin (at least, initially) rectangular strip, $L \times t$, in Fig. 14.3(b), whose initial orientation follows the angle θ inclined to the horizontal. The moment components in the local $(1, 2)$ symmetrical axes are $M_1 = M \cos \theta$ and $M_2 = M \sin \theta$, Fig. 14.3(c); correspondingly, $I_1 = Lt^3/12$ and $I_2 = tL^3/12$, which is larger than I_1 by $(L/t)^2$ (which can be much greater than unity when $L \gg t$).

Local curving of both (1-2) axes produces displacement components δ_1 and δ_2 normal to their original axes in the positive directions of M_1 and M_2, Fig. 14.3(c), respectively. If, for a straightforward calculation, we assume that all displacements are

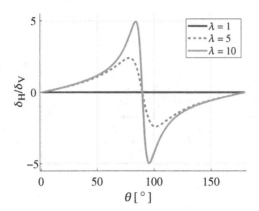

Figure 14.4 Variation of the ratio of lateral-to-vertical displacements for a rectangular section inclined at θ to the horizontal. Different thinness ratios are λ, with unity being square and larger values becoming narrower.

proportional to the level of local curving (as they would be if only external moments are applied), the ratio δ_1/δ_2 is $(M_1/M_2) \cdot (I_2/I_1)$ and thence $(L/t)^2 \cot \theta$.

From the vector diagram of displacements in Fig. 14.3(c), the total vertical displacement downwards is δ_V equal to $\delta_1 \cos \theta + \delta_2 \sin \theta$; to the left, and horizontally we have $\delta_H = \delta_1 \sin \theta - \delta_2 \cos \theta$. Their ratio, δ_H/δ_V, expresses the amount of sideways displacement compared to vertical. Conveniently, we can eliminate both δ_1 and δ_2 by dividing the top and bottom lines by δ_2 in order to substitute for our local ratio δ_1/δ_2. All δ terms naturally cancel, leaving us with:

$$\frac{\delta_H}{\delta_V} = \frac{\lambda^2 \cot \theta \sin \theta - \cos \theta}{\lambda^2 \cot \theta \cos \theta + \sin \theta}, \tag{14.2}$$

where $\lambda = L/t$. The performance is plotted in Fig. 14.4 for the indicated values of λ over a range of θ from zero to 180°, after which it repeats cyclically.

When λ equals unity, we have a square cross-section and not a strip, but the variation returns the zero line in Fig. 14.4. No matter how we rotate the square, displacements are only ever vertical under M. The behaviour mimics a circular cross-section in this regard but not absolutely because it can be shown that its vertical displacement depends on θ, being the same value every interval of 45° from $\theta = 0°$ onwards. Its bi-axial bending response, we say, is *centro-symmetrical*.

For values of λ greater than unity the strip becomes narrower: the displacement ratio increases, first positively then negatively on moving through $\theta = 90°$. The sign is not especially important and we expect a middle zero because the strip is purely vertical, preserving the displacement symmetry. The rate of change of behaviour increases as λ rises with narrower strips.

Horizontal displacements, however, begin to outpace those from vertical ($\delta_H/\delta_V > 1$) in the approximate range of 45° to 135° – when the distribution of strip cross-section is more vertical than horizontal, and increasingly so as λ rises. The sharp peak values (of around five for $\lambda = 10$) on either side of 90° are due to about one-sixth of M component-wise, underlining the dominance of the least-stiff axis and the exacerbation of displacements from increasing λ.

Finally, the snap-shot of our rotated strip in Fig. 14.3(b) expresses an asymmetrical cross-section with respect to the x- and y axes. It is clear that a single bending moment can produce dual-axis curving in the same way its components cause bi-axial bending about the local axes of symmetry. We should, therefore, expect misaligned displacements for asymmetrical cross-sections in general, likely to be driven by the weaker axis of bending (or curving).

14.3   Torsion

Analysis involving torsional stiffness is both similar and different to that for bending. An internal torque, T, is carried normal to the centroid of cross-section, whose deformation is a twisting rotation rate, ϕ, and no other in-plane distortion when warping is neglected.

T and ϕ are linked by the torsional stiffness of the cross-section which, like EI, is a product of material and geometrical constants: of G, the shear modulus, and J, the polar moment of area (or sometimes, torsional constant). Our generalised Hooke's Law for torsion is $T = GJ\phi$.

Calculating J for solid sections turns out to be non-trivial and beyond the scope here – unlike the relative ease when dealing with I and bending. Thin-walled sections are nevertheless amenable, with J given by the remarkable formula, with reference to Fig. 14.5:

$$J = \frac{4A_{\mathrm{E}}^2}{\oint \mathrm{d}s/t(s)}. \tag{14.3}$$

The cross-section does not have to be symmetrical nor convex – there can be inward facing edges. The through-thickness, t, is a function of a coordinate s moving intrinsically along the mid-thickness locus between the inner and outer wall surfaces, Fig. 14.5(a). We expect t to be small compared to any radial distance to it from the centroid G, and A_{E} is the *enclosed* area of cross-section, strictly measured to mid-thickness, see Fig. 14.5(b).

Evaluating J is generally easier compared to I. Apart from finding A_{E}, there is a single integration of the reciprocal of thickness around the cross-section, and if t is constant, $\oint \mathrm{d}s/t(s)$ is simply the ratio of arc-length to thickness.

For example, the cross-section of a type of laboratory scaffolding tube of constant thickness is shown in Fig. 14.5(c). The peripheral recesses permit other pieces of equipment to be mounted onto the tube, which inevitably apply offset forces and thus torques along the section. The shape, with its periodic concave recesses, is also known as a *re-entrant* cross-section.

The outer square dimension is $3a \times 3a$ with equal length recesses of size $a \times a$. The enclosed area is $9a^2 - 4a^2$ without finessing to mid-thickness (because $t \ll a$), and the total peripheral length is $20a$. From Eq. (14.3), $J = 4(5a^2)^2/(20a/t)$ equal to $5a^3t$.

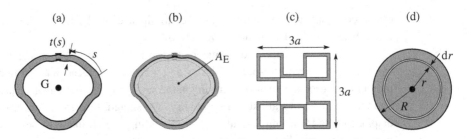

Figure 14.5 (a) General thin-walled hollow section subjected to a torque (not shown).
(b) Evaluation of area enclosed, A_E (and shaded), to mid-thickness. (c) Re-entrant, hollow
thin-walled cross-section. The internal corners only touch; they are not joined together.
(d) Solid circular tube cross-section.

A square tube of the same side-length without recesses has a J term equal to
$27a^3t$, almost five-and-a-half times larger. Recessing reduces the area enclosed as
well as increasing the peripheral length, both of which subtract from J; however, it is
necessary for this type of mounting application.

Out of interest, calculating the second moment of area about a horizontal axis (say)
is more involved because each plate element of length a has to be assessed individually
using the Parallel Axis Theorem. For simplicity, we can ignore I_G terms for horizontal
elements but not without several other terms:

$$
I_{xx} = 8\left[\frac{ta^3}{12} + a^2 \cdot at\right] + 4\left[\left(\frac{3a}{2}\right)^2 \cdot at\right] + 6\left[\left(\frac{a}{2}\right)^2 \cdot at\right] + 2\frac{ta^3}{12} = 19.33a^3t.
$$

$$(14.4)$$

For the same size of square tube, $I_{xx} = (3a)^4/12 - (3a - 2t)^4/12 \approx 18a^3t$. The
recessed shape now provides a moderate increase in bending stiffness, primarily from
increasing the number of plate elements contained within.

In general, 'placing' the wall of the cross-section as far away as possible from
the centroid improves both the torsional- and bending stiffness, and is expedited by
prismatic shapes such as circles and regular polygons. Bearing in mind the need for
a specific stiffness, we should compare sections of equal mass. For the same material
and wall thickness, this sets equal cross-sectional arc-lengths: if a circular tube has
radius R, a square tube has a side-length $a = \pi R/2$.

We know already that $I_{\text{circ}} = \pi R^3 t = I_{xx}$, and it is straightforward to show that
$J_{\text{circ}} = 2\pi R^3 t$, exactly twice. For the square tube $J_{\text{squ}} = a^3t$, giving us $(\pi^2/16)2\pi R^3 t$,
and I_{squ} can be shown to be $(\pi^2/12)\pi R^3 t$. The ratio of torsional constants in favour
of the circular tube is $16/\pi^2 = 1.62$, whereas for second moment of area we have
$12/\pi^2 = 1.22$. The circular tube is better in both.

The thin-walled circular tube is unique amongst shapes in that its torsional constant
can be used to infer that for thick-walled, even solid circular tubes. The underlying reason requires an equilibrium treatment we shall not consider. Other 'obvious'

cross-sectional shapes, such as a square and rectangle, do not have a well-defined J, and approximate expressions must be invoked.

For a solid tube, consider an elemental ring of area in Fig. 14.5(d) located at radius r and having 'thickness' dr. Its elemental torsional constant using Eq. (14.3) is $4(\pi r^2)^2/(2\pi r/dr)$, which is equal to $2\pi r^3 dr$. Integrating between limits of $r = 0$ and the outer radius R, we find $J = \pi R^4/2$. Note this value is also two times the second moment of area value.

The sizes of GJ and EI for any thin-walled section can now be compared informally. J can be up to twofold larger than I, and G is equal to $E/2(1 + \nu)$ for isotropic materials, which is around $0.38E$ for a typical Poisson's Ratio of 0.3. The ratio $(E/G) \cdot (I/J)$ is therefore around 1.25–1.4, giving a bending stiffness some 25%–40% larger.

14.4   Optimal Stiffness

Previously we compared I (and J) between circular and square tubes of the same mass by insisting on the same peripheral arc-lengths for equal thicknesses. This comparison of specific stiffness values can be extended to any cross-section in the following way.[1]

Its shape can be general, and we know its material area of bending (or torsion), A. We also know other properties such as I (about a given axis) and J, and these may have been calculated or measured, or read from a catalogue of available sections. Its area is the same as a 'standard' cross-section of our choosing, which happens to be a solid square in Fig. 14.6.

The second moment of area for the latter is $b^4/12$ about a horizontal axis, which can be re-written as $A^2/12$ and denoted as I_0. Comparing I for our cross-section of interest, we define a ratio I/I_0 now equal to $12I/A^2$. Since both areas are the same, by itself $12I/A^2$ expresses the performance of *all* cross-sections relative to the square.

A typical ratio value is around 60 for a familiar I-shaped cross-section made from steel. The thin-walled, $2a \times 2a \times t$ cross-section in Fig. 14.6(c) has an approximate wall area of $8at$ and, recall, $I_{xx} = 16a^3t/3$, giving a bending efficiency ratio, $12I/A^2 = a/t$. Being thin-walled, a/t is at least ten and much larger when the thickness approaches zero. Extremely thin walls, however, are susceptible to local buckling or crumpling and, thus, practical upper limits need to be set.

For torsion, we might compare against a circular solid cross-section because its torsional constant is known exactly, unlike the square. Defining $J_0 = \pi R^4/2$ with $A = \pi R^2$, the torsional efficiency ratio is $J/J_0 = 2\pi J/A^2$, where A is again the material area and not the area enclosed by the cross-section. As a quick example, a thin-walled circular tube has $J/J_0 = R/t$: again, much larger than unity.

We can plot these efficiency ratios on axes of A *vs* I (or J), in order to compare many cross-sections. For the same level of efficiency in either case, A is proportional

[1] P M Weaver and M F Ashby, The optimal selection of material and section-shape, *Journal of Engineering Design*, 7(2), pp. 129–150, 1996.

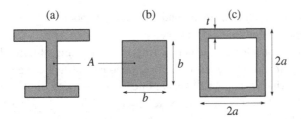

Figure 14.6 (a) General lightweight cross-section. (b) Solid square reference cross-section of the same area as (a). (c) Typical square hollow section.

to the square root of either I or J. Plotting points on logarithmic axes in A and I (or J), then additional lines with a gradient equal to 1/2 (from the square root) conform to the same efficiency but whose intercepts dictate different absolute values. Cross-sections with different efficiencies will, therefore, lie on different lines for an easier visual comparison.

Returning to our original question, what should the required stiffness of cross-section be, irrespective of whether it bends or twists, or both? Our material choice fixes E and G, often with little choice; having I and J as large as possible is ideal provided our section is not too heavy, which has to be borne with other applied loadings. Using the previous approach, we have a *provisional* sense of which shapes work best, but we have yet to quantify their properties.

Small displacements usually mean small rotations, but the reverse is not guaranteed. If we can bend a very long strip into an elastic circle and connect its ends, their displacements are very large even though local rotations (and curvatures) remain small. The *relative* scale of the problem is thus important, in this case being the ratio of initial length to strip thickness: we wish for a slender structure so that material yielding is not an issue but one that is not too slender that elastic deflections become excessive and flout the geometrical linearity of simple displacements.

Therefore, we cannot escape yielding considerations when enforcing small, elastic deflections. For example, consider a simply-supported beam carrying its self-weight as a uniformly distributed, vertical load. We want to ensure a reasonable span-to-depth ratio where displacements are small and elastic but just at their limit.

The density of beam material is ρ, giving a total self-weight $W = \rho g A L$, where g is the acceleration due to gravity. The central deflection, δ, is largest and equal to $5WL^3/384EI$ from Table 10.1. Replacing W and dividing by L for a sense of the relative deflection, we have $\delta/L = 5\rho g A L^3/384EI$.

Setting δ/L within a fractional inequality expresses an acceptable limit of relative deflection, which we denote as f: thus, $5\rho g A L^3/384EI < f$. A sense of the required slenderness is garnered from recognising that the second moment of area and the cross-sectional area itself are related by the radius of gyration, k, according to $I = Ak^2$. Substituting and re-arranging:

$$\frac{5\rho g A L^3}{384 E A k^2} < f \quad \rightarrow \quad \left(\frac{L}{k}\right)^2 < \frac{384 f}{5} \cdot \frac{E}{\rho g L}. \tag{14.5}$$

k tells us about the depth of cross-section about the bending axis, d, so L/k expresses the beam slenderness. The fractional limit f is often taken between 1/200 and 1/400 in practice, which sets the first right-hand side term, $384f/5$, to be around one-fifth. But $1/L$ is embedded in the second term where a large length – as in trying to bend our long strip into a circle – dictates a small slenderness, so that we cannot and, thus, avoid large displacements, and *vice versa*.

We prefer the right-hand side to be devoid of length and governed only by cross-sectional properties. Instead of substituting W for self-weight, we link W instead to material yielding in which the maximum bending moment in the middle, $WL/8$, just causes the outermost fibres to yield. From $\sigma = My/I$, the stress is σ_Y at a height $d/2$ above the neutral axis for a symmetrical cross-section. Consequently, $W = 16\sigma_Y I/dL$, and:

$$\frac{5}{384} \cdot \frac{16\sigma_Y I}{dL} \cdot \frac{L^2}{EI} < f \quad \rightarrow \quad \frac{L}{d} < \frac{384f}{80} \cdot \frac{E}{\sigma_Y}. \tag{14.6}$$

The first term on the right is around 1/50 to 1/100: the second term is approximately 350 for aluminium and 750 for steel. The length-to-depth ratio can therefore be anywhere between 7 and 15, which again is the usual practice. For larger ratios, the beam is relatively longer and the deflection increases. As such, its performance is now *stiffness limited* rather than by yielding. Shorter beams will tend to yield before the deflection limit is reached.

The question of the required stiffness is implied by having small displacements, but we can make I explicit by taking a general cross-section and writing the maximum moment in terms of the *section modulus, Z, i.e. $Z \cdot \sigma_Y$*. When material yielding first occurs, Z is the *elastic* section modulus; it is the *plastic* section modulus when the cross-section yields entirely, but this is not our concern – both are dealt with in the following chapter. Our limiting slenderness formula is modified to become:

$$\frac{L}{d} < \frac{384f}{40} \cdot \frac{E}{\sigma_Y} \cdot \frac{I}{Zd} \quad \rightarrow \quad \frac{I}{Zd} > \frac{L}{d} \cdot \frac{40}{384f} \cdot \frac{\sigma_Y}{E}. \tag{14.7}$$

Using either formula depends on whether we declare I/Zd or L/d a priori. Either way, I is not an independent parameter but embedded always within the ratio I/Zd. This should not be surprising given that yielding has to be part of our analysis, but that our analysis is also general; we are not solving for a specific geometry or load set, and a relative measure is always output.

Interestingly, the ratio I/Zd can be shown to be less than one-half for any cross-section – think of an extreme I-section where the area is exactly concentrated in equal halves at flanges d apart. In the limit of one-half, we therefore arrive back at Eq. (14.6).

In closing, designing for optimal stiffness is dictated by relative performance measures if we are seeking insight as closed-form expressions. Without doubt, there are more metrics to consider in practice: multiple loads, separate structural actions such as torsion and shearing, as well as multiple members and their joints.

These *multiobjective* problems are tackled successfully by using numerical algorithms which 'search' for solutions that offer a compromise for competing metrics.

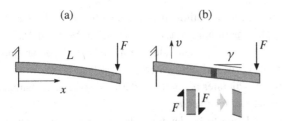

Figure 14.7 (a) Tip-loaded uniform cantilever. (b) Shearing deformation of beam element.

We therefore submit to the numerics of the search process, which can subtract from physical insight, but often a sensible starting solution is required to work forward, and the methods above are eminently suitable.

14.5   Final Remarks

When slender members bend, transverse displacements are governed by curving of the beam centre-line, we are often told. The contribution from shearing can be assessed quickly for beams with straightforward cross-sections and concludes the same for more general shapes.

Our final example is a cantilever with a solid cross-section, $b \times d$, and loaded by a tip force, F, Fig. 14.7(a). The well-known tip deflection downwards from bending alone is $FL^3/3EI$, equal to $4FL^3/Ebd^3$ after replacing I.

There is a constant shear force, F, throughout, which tends to distort any narrow rectangular element of the beam in side-view into a gentle parallelogram, Fig. 14.7(b). The change in angle at any original corner measures the shear strain, which is denoted as γ. This is also the same angle as the current gradient of the beam centre-line, given by dv/dx for transverse displacements, v.

Despite a constant shear force at every cross-section, the shear stress, however, varies from top to bottom in a quadratic fashion: we do not cover the details here but note that it ranges from zero to a maximum value on the centre-line equal to 3/2 times the 'average' shear stress F/bd, and back to zero.

From the centre-line value, we can express the shear strain in terms of F via $G\gamma = 3F/2bd$, and ultimately v after replacing γ and integrating. With no rotation at $x = 0$ due to the built-in end, we find $v = 3Fx/2Gbd$ and, thus, the component of shear deflection at the tip.

The total tip deflection combines the sizes of both components $4FL^3/Ebd^3 + 3FL/2Gbd$, or in relative terms with respect to the bending displacement:

$$\frac{4FL^3}{Ebh^3}\left[1 + \frac{3E}{8G}\frac{d^2}{L^2}\right]. \tag{14.8}$$

The term $3E/8G$ is around unity for typical materials and d/L is of the order of one-tenth given the previous analysis for slender beams. The second term inside brackets is thus very small, around one percent, if at all: the component of shear displacement is, therefore, negligible.

15 Cross-Sectional Strength

The strength of a material is defined most simply by its yield limit. For ductile materials, this occurs at highly repeatable values of both stress and strain; for others, such as brittle materials, their strength can vary widely, depending on the size of the specimen.

Ductility also guarantees a capacity *for* yielding: the material can stretch over much larger strains, often at higher stresses before ultimately failing by tearing. Engineers usefully simplify this response as two linear regimes, in tension and compression, see Fig. 15.1.

The initial elastic phase has a familiar gradient, E, equal to the Young's Modulus before terminating at a yield stress $\sigma = \sigma_Y$; the corresponding yield strain is $\epsilon_Y = \sigma_Y/E$. Following this there is a shallower phase – usually a plateau, for plastic straining at constant stress up to some nominal failure strain. Unloading from the plastic phase is immediately elastic, as shown.

Because the actual transition between phases is gradual, it is also customary to apportion ϵ_Y as a nominal plastic strain value after unloading, typically 0.2% for metals. The corresponding value of plastic stress somewhere above initial yield is now designated the yield stress, but we shall keep to the first, simpler bilinear model.

Yielding of a *structure* depends on which cross-section is most heavily loaded and, within that, where the stresses are highest. A bar in uniform tension or compression yields everywhere at the same time, well before the necking instability concentrates plastic deformation at much higher strains. The strength in tension or compression (provided axial buckling is not a problem) is therefore measured by the product of cross-sectional area, A, and σ_Y, which must be greater than the bar force, *c.f.* Chapter 6.

At the cross-section of the largest beam bending moment, M, material yielding begins at the top or bottom when the peak linear elastic stress reaches $\pm\sigma_Y$ according to $\sigma = M(\pm y_{max})/I$. For simplicity, we consider cross-sections that are vertically symmetrical so that I is the corresponding second moment of area in the direction of M: otherwise, there is coupling between M and lateral bending and its I. In Fig. 15.2(a), the bottom-most set of fibres have just yielded for a typical T-section; y_{max} for the top fibres is lower, which remain elastic.

The onset of this critical bending can be defined by setting $M = M_Y$, the *first yield moment*. The stress formula can be inverted to give $M_Y = (I/y_{max}) \cdot \sigma_Y$, and

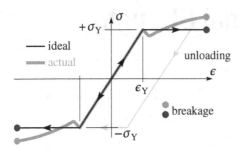

Figure 15.1 Stress *vs* strain response for a typical ductile material. The black bilinear curve idealises the true behaviour (grey).

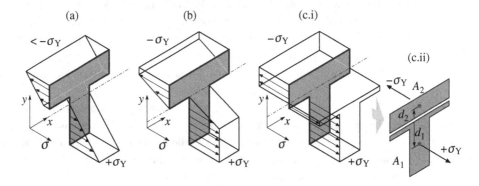

Figure 15.2 Progressive yielding of a T-section due to bending about x-axis. (a) Linearly elastic axial stresses with the onset of material yielding in the bottom-most fibres. (b) Yielding in the top and bottom parts with an elastic core. (c.i) Limiting fully plastic behaviour; (c.ii) areas of tensile and compressive yield stresses.

abbreviated to $Z_E \cdot \sigma_Y$. The subscript 'E' denotes an elastic limit for the cross-section, and Z_E is a purely geometrical term known as the *elastic section modulus* (*c.f.* Chapter 14).

As the bending moment rises, the stresses saturate at $\pm \sigma_Y$ for plastic strains beyond $\pm \epsilon_Y$, which maintain a linear height-wise distribution for deformed plane cross-sections to remain planar normal to the axis of bending. Yielding begins on both sides of the neutral axis with the plastic regions steadily encroaching into the linear elastic stress distribution, Fig. 15.2(b), to give an elasto-plastic mixture of stresses.

The limiting response occurs when all stresses reach either $+\sigma_Y$ or $-\sigma_Y$, Fig. 15.2(c.i), with M now equal to M_P, *the fully plastic moment* and ultimate capacity in bending. Before calculating its value, we note that the neutral axis divides a pair of uniformly stressed regions, A_1 and A_2, Fig. 15.2(c.ii).

Multiplying each area by either $\pm \sigma_Y$ gives a pair of centroidal axial forces located at distances d_1 and d_2 below and above the axis, respectively, with our now sagging M_P equal to $A_1 d_1 \sigma_Y + A_2(-d_2) \cdot (-\sigma_Y)$. No axial force is applied to the beam,

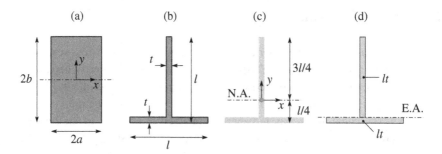

Figure 15.3 (a) Solid rectangular section. (b) Inverted thin-walled T-section. (c) Location of centroid for (b), assuming $l \gg t$. (d) Location of equal-areas axis.

setting A_1 equal to A_2, and the neutral axis as the *equal-areas* axis. For uni-axial bending of unsymmetrical cross-sections in general, this axis differs from the original elastic neutral axis, which 'migrates' during yielding.

It is often easier to sub-divide the cross-section about the equal-areas axes into more than two, simpler, prismatic shapes of areas A_r. With their centroids located at normal distances d_r, M_P is a simple summation equal to $\sum A_r |d_r| \sigma_Y$ for all terms being positive. Analogous to M_Y, we can write $M_P = Z_P \cdot \sigma_Y$, with the *plastic section modulus* defined by $Z_P = \sum A_r |d_r|$.

Two example cross-sections carry a horizontal bending moment in Fig. 15.3: a solid rectangle of $2a \times 2b$ in Fig. 15.3(a) and a thin-walled inverted T-section of equal length legs, l, with thickness t, Fig. 15.3(b).

Recall $I = 2a(2b)^3/12 \, (= I_{xx})$ for the rectangle, giving $Z_E = (16ab^3/12)/b = 4ab^2/3$. The equal-areas axis symmetrically divides the cross-section, setting $Z_P = 2 \times 2ab(b/2) = 2ab^2$, which is larger than the elastic value by half as much again; the bending moment beyond the first yield moment and, possibly, the loading can increase by 50% before fully yielding.

If we assume that t is much smaller than l for the T-section, its centroid is located $l/4$ above the base line, Fig. 15.3(c). The second moment of area is calculated via the Parallel Axis Theorem, neglecting I_G about the centroid of the base flange because it depends on t^3. As a result:

$$I = \frac{t(3l/4)^3}{12} + \frac{3lt}{4}\left(\frac{3l}{8}\right)^2 + lt\left(\frac{l}{4}\right)^2 = \frac{52l^3t}{256} \approx 0.20l^3t. \qquad (15.1)$$

The furthest fibre lies at $3l/4$ above the neutral axis, giving $Z_E = I/y_{\max} = (52l^3t/256)/(3l/4) = 52l^2t/192 \approx 0.27l^2t$.

The equal-areas condition now sets its axis at the junction of the two legs, Fig. 15.3(d), moving away from the original neutral axis with lt on either side. Notwithstanding, $Z_P = lt(l/2) + lt(t/2) = (l^2t/2) \cdot (1 + t/l)$, and if we neglect higher order terms in t, $Z_P \approx l^2t/2$, which is almost twice as large as Z_E.

15.1   Plastic Moments

More interesting behaviour emerges when more than one stress-resultant is present at the critical cross-section. Elastic bi-axial bending in Chapter 14 had a different neutral axis compared to the direction of the bending moment resultant, Fig. 14.3, so let us consider plastic bi-axial bending of the same solid cross-section.[1]

The equal-areas axis is generally inclined at angle γ to the horizonal (and differently from θ, we assume, the direction of the resultant bending moment, M) in Fig. 15.4(a).

From rectangular symmetry, the axis passes through the centroid of cross-section, intersecting vertically at $\beta_1 b$ above the centroid, Fig. 15.4(b). Beyond $\beta_1 = 1$, the axis meets the horizontal edges, which requires a separate calculation later in terms of a second variable, β_2. For now, γ is a positive angle less than $\arctan(b/a)$.

Each area furnishes equal and opposite axial yield forces after multiplying by σ_Y, as before. The resulting moment, however, has to be related to M and, because γ and θ are different, M exerts a component in the direction of the γ axis. Furthermore, if the centroid of forces on each half normal to the equal-areas axis are not collinear, they also exert another moment component about a new axis, perpendicular to this one. Calculating the net critical bending moment in terms of M, θ and γ can very quickly become wieldy.

More straightforwardly, we consider the original components $M_x = M \cos \theta$ and $M_y = M \sin \theta$. Each equal area is then sub-divided into regular, smaller areas with respect to the original (x, y) system giving simpler expressions for their centroidal distances from both axes. Here, we have four regions for each positively (p) or negatively (n) stressed half: two rectangles, p_1 and p_4, and two triangles, p_2 and p_3, and their four negative counterparts $n_1 \ldots n_4$.

Simply finding the sum of $A_r |d_r|$ for Z_P about both axes, however, will not work, because there are now equally stressed regions about either. Absolute values of d_r and yield stress are required in order to establish the sign of their product, which are shown in Figs. 15.4(c) and (d) for bending about the x- and y-axis, respectively. In particular, for M_x we have:

$$M_x = \underbrace{2 \times ab \cdot \frac{b}{2} \cdot \sigma_Y}_{p_1, n_1} + \underbrace{2 \times ab(1 - \beta_1) \cdot \frac{b(1 + \beta_1)}{2} \cdot \sigma_Y}_{p_4, n_4}$$

$$+ \underbrace{2 \times \frac{1}{2} a\beta_1 b \cdot \frac{2}{3}\beta_1 b \cdot \sigma_Y}_{p_3, n_3} - \underbrace{2 \times \frac{1}{2} a\beta_1 b \cdot \frac{1}{3}\beta_1 b \cdot \sigma_Y}_{p_2, n_2}. \qquad (15.2)$$

After simplifying and re-arranging:

$$M_x = 2ab^2(1 - \beta_1^2/3) \cdot \sigma_Y. \qquad (15.3)$$

---

[1] J Heyman, *Elements of the Theory of Structures*, Ch. 2., Cambridge University Press, 1996.

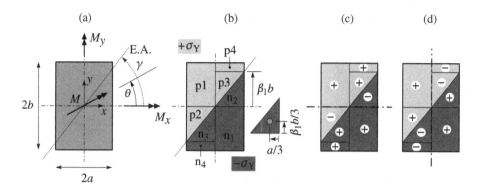

Figure 15.4 (a) Plastic bi-axial bending of a solid rectangular cross-section by a moment inclined at θ to the horizontal; the equal-areas axis is inclined at γ to the same. (b) Sub-division of tensile and compressive yield stress regions either side of the equal-areas axis into simpler shapes. The axis intersects the vertical sides only. (c) Numerical sign $(+/-)$ for the product of yield stress and centroidal position with respect to the horizontal axis. (d) The same product now with respect to the vertical axis.

For M_y, we similarly write using Fig. 15.4(d):

$$M_y = \underbrace{2 \times ab \cdot \frac{a}{2} \cdot \sigma_Y}_{p_1, n_1} - \underbrace{2 \times ab(1 - \beta_1) \cdot \frac{a}{2} \cdot \sigma_Y}_{p_4, n_4}$$

$$- \underbrace{2 \times \frac{1}{2}a\beta_1 b \cdot \frac{a}{3} \cdot \sigma_Y}_{p_3, n_3} + \underbrace{2 \times \frac{1}{2}a\beta_1 b \cdot \frac{2}{3}a \cdot \sigma_Y}_{p_2, n_2} = a^2 b \frac{4\beta_1}{3} \cdot \sigma_Y. \quad (15.4)$$

Remembering that $\tan \theta = M_y/M_x$, we find after substitution, $\tan \theta = (2/3) \cdot (a/b)\beta_1/(1 - \beta_1^2/3)$. Then, replacing β_1 via $\tan \gamma = \beta_1 b/a$ from Fig. 15.4(a), a quadratic equation emerges:

$$\tan^2 \gamma \tan \theta + 2 \tan \gamma - 3(b/a)^2 \tan \theta = 0 \quad (15.5)$$

whose roots are $\tan \gamma = -1 \pm (1 + 3(b/a)^2 \tan^2 \theta)^{1/2}$.

The larger one is correct and quite different from $\tan \theta$, even when θ is very small, signifying almost uni-axial bending with $\tan \gamma \approx (3/2) \cdot (b/a)^2 \tan^2 \theta$; the other smaller root in this case is approximately equal to -2, which is physically invalid. Recall from elastic bi-axial bending in Chapter 14 $\tan \gamma = (b/a)^2 \tan \theta$.

A steeper equal-areas axis intersects the horizontal edge now at $\beta_2 a$ normal to the y-axis, Fig. 15.5(a), revealing:

$$M_x = 4ab^2(\beta_2/3) \cdot \sigma_Y, \quad M_y = 2a^2 b(1 - \beta_2^2/3) \cdot \sigma_Y \quad (15.6)$$

with $\tan \gamma = (b/a)/\beta_2$, which may be expressed in terms of θ, if desired.

Rather than specify M by its size and direction, we work 'towards' its performance by first setting β_1 and β_2 in the range zero to unity in order to calculate M_x, M_y and γ; the arc-tangent of M_y/M_x sets θ. Moment values are positive with a specific

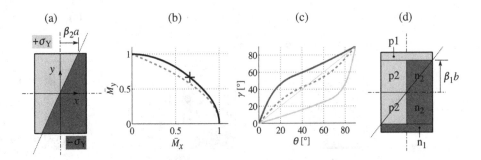

Figure 15.5 (a) Extension of plastic bi-axial model from Fig. 15.4 where equal-areas axis intersects horizontal edges. (b) Combined limit of plastic moment components. The black curve is for exact yield stresses; the grey curve relates to the approximate distribution in (d). The cross is where the equal-areas axis intersects the corners of the cross-section. (c) Variation of equal-areas inclination with angle of overall moment for $b/a = 0.5$ (light grey), $b/a = 1$ (grey) and $b/a = 1.5$ (dark grey). The fine straight line is a construction. (d) Approximate yield stress distribution using rectangular stress blocks only.

first-quadrant relationship when plotted in Fig. 15.5(b). Negative values of β_1 and β_2 $(-1 \rightarrow 0)$ allow the equal-areas axis to be rotated 'fully' around the cross-section, if required, giving three more quadrants.

For absolute values of M_x and M_y, the resultant moment, M, is the radial distance from the origin, with θ as its inclination. However, dimensionless components are plotted for universality, by dividing by σ_Y and their equivalent values of Z_P for uni-axial bending, for example, about the y-axis, $Z_P = 2b(2a)^2/4$, with $\bar{M}_y = M_y/(2ba^2\sigma_Y)$. A small cross marks the changeover from $\beta_1 = 1$ to $\beta_2 = 1$.

We plot the variation of γ with θ in Fig. 15.5(c) for three values of b/a, from a flat to a taller rectangle through square: 0.5 (light), 1.0 (grey) and 1.5 (dark). The lowest value suggests that γ 'lags' θ by increasing more slowly at first before a final rising flourish; the behaviour reverses for the largest value. The response, however, for $b = a$ shows both lagging and leading about the line $\gamma = \theta$. All variations recombine at $\theta = 0°$ and $90°$, the two points of uni-axial bending.

A second, grey curve is plotted in Fig. 15.5(b) for a different distribution of yield stresses shown in Fig. 15.5(d). Their layout is parameterised using the same β_1 but the equal-areas axis is now a staggered line separating rectangular sub-areas. This 'coarser' distribution of stress eliminates the previous triangular regions in favour of a simpler calculation of bending moment parts, and its anti-symmetrical nature contributes to both M_x and M_y; in particular, regions p_1 and n_1 relate to M_x, and p_2 and n_2 to M_y, whence:

$$M_x = 2ab^2(1 - \beta_1^2) \cdot \sigma_Y, \quad M_y = 2a^2b\beta_1 \cdot \sigma_Y. \tag{15.7}$$

Setting β_1 from zero to unity sufficiently captures first-quadrant behaviour between M_x and M_y, and their dimensionless curve lies inside the previous exact case, signifying a lower estimate of the critical moment value.

If we adopt the latter as our critical bending model, the true strength of the cross-section will be marginally greater. This is precisely a *Lower Bound* solution, now applied to the cross-sectional capacity; correlating this to the responsible external loads would be a natural next step.

In return, we have a simpler calculation from the rectangular geometry alone. Of course, the exact calculation is not much more difficult, but we are exercising the principle for dealing with, next, combinations of different stress resultants.

15.2 Plastic Moment and Force

The rectangular cross-section in Fig. 15.6(a) carries a single, horizontal bending moment, M, and an axial force, T, in tension. Equal and opposite blocks of yield stress with cross-sectional area A_1 maximise the moment leverage about a central block of area $A_2 = 2a \cdot 2b\beta$ dedicated to 'furnishing' T alone. Clearly, $T = 4ab\beta\sigma_Y$, and taking moments about the mid-depth axis sets M equal to $2ab^2(1 - \beta^2) \cdot \sigma_Y$

The largest force occurs when $\beta = 1$ with $T = 4ab\sigma_Y$; conversely, when $\beta = 0$, the stresses are symmetrical and furnish a pure moment, $M = 2ab^2 \cdot \sigma_Y$. Dividing each general expression by these respective limits and denoting by an over-bar, we arrive at $\bar{T} = \beta$ and $\bar{M} = 1 - \beta^2$: or $\bar{M} + \bar{T}^2 = 1$.

This is the *interaction equation* between the limiting combination of force and moment, which can be plotted, as before, now on axes of $\bar{M}$ vs $\bar{T}$. The coordinates of bending moment and axial force from the actual loading must lie inside this curve for sufficient strength as predicated by the Lower Bound Theorem; if they lie on the curve, the cross-section is doubly critical, and outside the curve, it surely fails.

Next, T is replaced by a shear force, S, Fig. 15.6(c). This similarly addresses two distinct stress resultants, with S governed by the yield limit in shear, τ_Y, acting over a central core to preserve maximal moment leverage: ascribing symmetrical stress block areas as before using β, we would find the same interaction equation between M and S as between M and T (for suitable dimensionless groups).

Shearing on first sight is different to axial force behaviour, and both of their stresses can be active over the same area – as known from elastic behaviour. Some of the central core responsible for S alone can, therefore, be devoted to extra moment capacity by introducing two more axial stress blocks as shown in Fig. 15.6(d); before, axial stresses had to be devoted to T or to M alone.

At a material level, however, axial stresses, σ, and shear stresses, τ, cannot both be at their respective yield levels; there is a local limiting interaction equal to $\sigma^2 + 3\tau^2 = \sigma_Y^2$ from von Mises.[2] Shearing and axial stresses are, in fact, physically correlated during yielding, and the von Mises equation is a valid Lower Bound statement of their equilibrium.

---

[2] C R Calladine, *Plasticity for Engineers, Theory and Applications*, Woodhead Publishing, 2000.

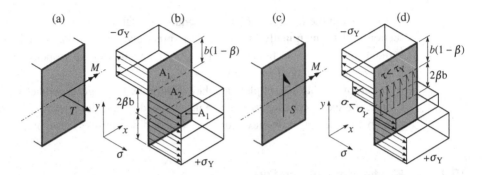

Figure 15.6 (a) Plastic bending and extension of a rectangular solid cross-section.
(b) Yield-stress distribution: the middle core applies only to the tension force, the outer regions for moment. (c) Plastic bending and shearing of the same cross-section. (d) Corresponding stress blocks where the middle core carries both ordinary- and shear stresses in relation to the von Mises equation.

Assuming a nominal proportion of axial stress in the central core defined by $\sigma = \lambda\sigma_Y$, it is easy to show that the moment capacity can be increased by $2 \cdot 2a\beta b \cdot (\beta b/2) \cdot \lambda\sigma_Y$. The corresponding value of shear force is $S = 2a \cdot 2\beta b \cdot \tau$.

The limit of shear yielding according to the von Mises equation has $\tau = \tau_Y$ equal to $\sigma_Y/\sqrt{3}$ when $\sigma = 0$. We can, therefore, write for both stress resultants:

$$\bar{M} = \frac{M}{2ab^2\sigma_Y} = 1 - \beta^2 + \beta\lambda, \quad \bar{S} = \frac{S}{4ab\tau_Y} = \beta(1 - \lambda^2). \tag{15.8}$$

The interaction between moment and shear force now depends on two parameters that can be independently adjusted between zero and unity. There is extra freedom to tailor critical values, but ultimately the moment increases at the expense of a reduction in S. The augmented model, nonetheless, shows how we might broaden our design scope in general.

15.3   Plastic Connections

In two final examples we show how a Lower Bound equilibrium approach can be extended to basic *connection* design. Although similar in principle, a connection is prone to several failure modes because they contain more structural elements by their nature; there is thus scope for multiple interactions and optimisation between these modes. It is usual, however, to assume the simplest equilibrium solution for each mode in order to over-estimate the complete capacity, and generously so, as connections can be the most critical link in any structure.

First, a rectangular bar, $a \times b$, is rigidly built into a wall in Fig. 15.7(a). Close to the tip, a total force of T is transmitted as a pair of equal axial forces, $T/2$, applied to the ends of a cylindrical barrel, which lies perfectly inside a machined hole of diameter ϕ. We wish to understand the transmission limit of the bar and barrel.

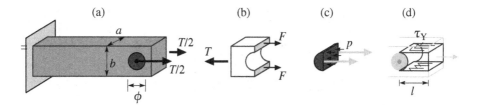

Figure 15.7 (a) Axially loaded rectangular bar by pulling on a barrel inserted into a hole. (b) Tensile failure around the hole. (c) Bearing failure of the barrel. (d) Shearing failure ahead of the barrel.

Well away from the barrel end, T must be less than $T_1 = ab \cdot \sigma_Y$ if the bar itself does not yield in tension. The cross-sectional area, however, diminishes moving along the barrel, being minimal at full diameter, Fig. 15.7(b). The tension divides equally into a pair of forces, F, each carried over an area of $a(b - \phi)/2$, giving a limiting tension $T_2 = a(b - \phi) \cdot \sigma_Y \leq T_1$.

The face of the barrel tip-side is being compressed against the inside of the hole, leading to a *bearing* pressure, p, between the two surfaces. The distribution of pressure varies from top-to-bottom, but for a uniform and thence approximate distribution acting over the projected area, $a\phi$, of the hole as shown, and operating at yield, i.e. $p = \sigma_Y$, the tension must be less than $T_3 = a\phi \cdot \sigma_Y$.

The lowest possible tension 'governs' failure, and T_1 is clearly remiss. If we can specify a given hole size, T_2 and T_3 become equal when $\phi = b/2$. This would suggest simultaneous failure in both modes, although in practice they do not because each equilibrium solution is over-simplified.

A final mode involves the barrel pulling through the end of the bar and shearing a plug of material of length, l, Fig. 15.7(d). There are two shear 'planes' on which the shear yield stress, τ_Y, is present, setting the tension to be less than $T_4 = 2al \cdot \tau_Y$. Taking $\tau_Y \approx \sigma_Y/\sqrt{3}$, we can find a limit for l at which T_4 becomes equal in value to the other modes. Typically, l is smaller than b, so the barrel pulling out this way occurs only if it is very close to the tip.

The bar is now connected to the base through a pair of right angled 'cleats' in Fig. 15.8(a), which are glued to the top and bottom surfaces. Each cleat is then bolted to the wall at a single spot a height d above and below the bar surfaces. The bar also carries an end moment, M, at its tip.

There are many possible failure modes, but we shall consider only a few for the cleats themselves. The bolt axes are separated by $b + 2d$, see Fig. 15.8(b), where equal and opposite bolt forces, T, generate a couple equal to M. In order to avoid yielding of the bolt shanks, each cross-sectional area must be greater than $M/([b + 2d] \cdot \sigma_Y)$.

Each bolt force is transmitted through the cleat to the glued surfaces, which equilibrate under shearing forces, S, Fig. 15.8(c). The corresponding shear stress τ, we assume, is uniformly distributed over the area, which depends on the leg length of cleat, l, with $\tau \approx S/(la) \leq \tau_Y$, which sets a minimum value of l for

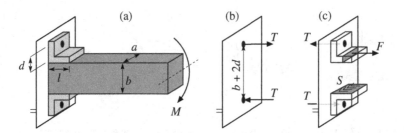

Figure 15.8 (a) Bending of a cantilever glued to right-angled cleats, themselves bolted to a vertical wall. (b) Tensile bar forces at yield generate a limiting moment at the base. (c) Limiting axial and shear forces on the horizontal legs of each cleat.

$S = T = M/(b + 2d)$. Likewise, the leg thickness must be large enough to give normal stresses, F/at, smaller than σ_Y, with F approximately equal to T.

A complete design should address the upright part of each cleat as well as their local moment equilibrium in more detail; that $F = T$ in Fig. 15.8(c) neglects any other forces necessary for moment equilibrium (when looking at each cleat from the side). Suffice to say, $F = T$ is the worst case scenario for assessing the required leg thickness when we do consider moments, which is fortunate.

15.4   Final Remarks

Provided our equilibrium solution affords critical stresses that nowhere exceed yielding levels, we can obtain an estimate of the strength of cross-section in terms of limiting stress resultant values; by not exceeding yielding we mean, of course, that up to everywhere within the cross-section, the material may be just about to yield.

When more than one stress resultant is present, the capacity of the cross-section in all is naturally reduced. Their critical interaction can be ascertained with precision if we choose or, by a generous interpretation of the Lower Bound Theorem, we can simplify this calculation at the expense of a reduced capacity by choosing an approximate but viable equilibrium solution. This is especially pertinent for connector design, whose multiple failure modes depend on the local geometry as well as the local stress resultants.

The interaction between two resultants in general can be expressed graphically as a bounding curve outside of which their combination promotes failure. Inside, the cross-section will be able to operate without yielding fully – as will any approximate solution.

16   Back of the Envelope: Beam Design by Lower Bound

One of the originators of structural plasticity, J F Baker, continued to favour his Lower and Upper Bound methods during the 1950s as computer technology was emerging for numerical structural analysis: structural design should not be governed by exact solutions alone but also by a healthy appreciation of viable equilibrium solutions – so-called 'load paths', of making construction materials behave in a ductile fashion, and of load redistribution when material yielding commences.

Baker was correct. But with bespoke and cost-effective thin-walled steel members more liable to buckling than yielding, exact stresses must be imputed to modern codes of practical design. It is then a trivial step with current analysis software to calculate all of the structure's performance exactly.

There is still a place for design by Baker's plasticity methods: they are quick – amenable to back-of-the-envelope scribbles, they gird our expectations for the scale of the exact solution and they are always conservative (although buckling is not accounted for).

We focus here on Lower Bound design for simple yet indeterminate beams whose elastic performances quickly succumb to formidable algebra. We are not interested in optimising their load-carrying capacity, where the best safe load 'rises' to meet the lowest collapse load from an Upper Bound approach.

Rather, the role of supports and other types of structural members tends to be overlooked during such studies, usually in favour of continuous multi-span beams and how to make them better by adjusting the size of certain beam sections.

We discussed in Chapter 6, for example, the rarely reported Lower Bound performance of a truss; what about structures having a mixture of elements, such as bars and beams? Supports are expected to furnish the necessary boundary conditions for enabling the safe redistribution of bending moments: what if this was not so obvious?

First, we highlight the efficacy of the method compared to elastic analysis for a propped cantilever: as per the truss from Chapter 6, the reduction in analytical effort is writ large.

We then consider three designs of a 'long' beam where a typical simple support is replaced first by a T-junction of the type described in Chapter 12, restraining the beam vertically but rotating against the bending of the new support leg. The support leg is then removed altogether and replaced by a tension tie, thereby mixing the structural make-up.

16.1   The Method

Figure 16.1(a) shows a cantilever beam of length L and bending stiffness EI, built into a rigid wall at one end and simply supported at its tip. A vertical, central point force, W, is applied from zero up to plastic failure. The beam material is ductile, and we set the built-in bending moment as the single redundancy.

The elastic solution combines the 'free' case – from releasing this moment by inserting a pin or converting to a simple support, Fig. 16.1(b) – before reinstating it alone without W for the 'reactant' case, Fig. 16.1(c). Hogging bending moments are taken to be positive, and we distinguish the salient (peak) moment values by M_1 and M_2, as shown, the latter being the reactant. Point forces throughout dictate piece-wise linear profiles.

The end rotations in each case are standard results from Chapter 10 and must sum to zero:

$$\theta_1 = \frac{WL^2}{16EI}, \quad \theta_2 = \frac{M_2 L}{3EI} \quad \rightarrow \quad M_2 = \frac{6WL}{32}. \tag{16.1}$$

It is then trivial to show that $M_1 = -5WL/32$, given vertical reactions of $5W/16$ and $11W/16$.

The ratio $|M_1/M_2| = 5/6$ ensures that the built-in bending moment is the first to 'reach' the yield moment, M_Y, and then the fully plastic moment, M_P, as W increases. Depending on the shape of the cross-section, $|M_1|$ may have already exceeded M_Y (but not reached M_P) when $M_2 = M_P$.

By this point, a plastic hinge has formed at the built-in end, where the rotation is no longer zero. Our earlier compatibility condition is not upheld, but the structure still stands and W is safely carried. It is only when M_1 becomes equal in size to M_P that a second hinge forms, leading to collapse: at a fraction before this point, the largest possible load is *just* safely carried.

Its magnitude can be distilled from an exact elasto-plastic analysis; or as the Lower Bound Theorem requires, from an equilibrium solution alone that nowhere violates the yielding condition. Since bending dominates the equilibrium response, moments everywhere have to be less than M_P in size.

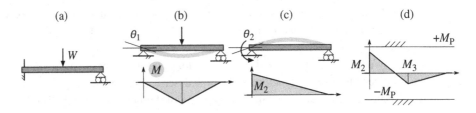

Figure 16.1 (a) Centrally loaded, propped cantilever. (b) Free response after releasing the built-in support, including bending moment profile. (c) Reactant response after reinstating built-in moment. (d) General bending moment profile sitting between fully plastic limits.

In theory, there can be many equilibrium solutions because compatibility is not considered; how much the first hinge above has rotated does not matter. But it will be satisfied for the best equilibrium choice, which just equals the lowest collapse load.

Correspondingly, salient bending moments are often set to be fully plastic values, requiring the final profile in Fig. 16.1(d) to glance 'hard' limits of $\pm M_P$. As such, $M_2 = +M_P$ and M_1 is equal to $-WL/4 + M_P/2$, which is assumed to be negative in total despite a positive reactant contribution.

Setting M_1 equal to $-M_P$, however, obscures a valuable lesson in how the inequality works (and often a source of algebraic error); as a negative total, M_1 must also be *greater* than $-M_P$, giving $-WL/4 + M_P/2 \geq -M_P$. After rearrangement in terms of W, then $W \leq 6M_P/L$, with the inequality correctly indicating the largest safe limit of loading.

There is no reason, as is often insisted by teachers, to maximise salient values to their plastic limits. Of course, this gives the best performance for the structure, but if we did not then a curious oddity of theoretical behaviour emerges compared to the actual – as we saw in Chapter 6. For example, let us set a smaller value of M_2 equal to βM_P, where β is less than or equal to unity in size. The inequality statement for M_1 now affords $W < (4M_P/L) \cdot (1 + \beta/2)$.

The original result is naturally given by $\beta = 1$. When $\beta = 1/2$, then W is a smaller value, $5M_P/L$. Remarkably, when $\beta = 0$, which suggests no built-in moment and thus the equivalent of a pinned support, W is reduced again to $4M_P/L$. Even when $\beta = -1$, signifying that the built-in bending moment has (magically) reversed (where yielding is still not strictly violated), $W = 2M_P/L$.

A valid equilibrium solution is all that matters. We do not have to satisfy the rotational boundary conditions provided mechanistic behaviour is not incurred: the last two interpretations of β values show this clearly. The corresponding safe load is always lower than what is actually possible, and if our choice of equilibrium solution seems perplexing in the real world but yields a simpler calculation, we might embrace it.

Before moving on to the long beam, we might ask about shear forces and their tendency to reduce the fully plastic moment capacity of a given cross-section. As shown in Chapter 15, their interaction was defined by a pair of parametrical equations normalised by their respective fully plastic limits, and for a solid rectangular cross-section, $2a \times 2b$, these were shown to be:

$$\bar{M} = \frac{M}{2ab^2\sigma_Y} = 1 - \beta^2, \quad \bar{S} = \frac{S}{4ab\tau_Y} = \beta; \quad \bar{M} + \bar{S}^2 = 1. \tag{16.2}$$

The largest shear force is in the first half of the beam and equal to $2M_P/(L/2)$ from the bending moment gradient in Fig. 16.1(d): both salient moments are thus affected. For this cross-section, the shear force can be written as $S = 4M_P/L = 8ab^2\sigma_Y/L$, giving $\bar{S} = 2(b/L) \cdot (\sigma_Y/\tau_Y)$.

The final yield stress ratio is (approximately) $\sqrt{3}$, so we find $\bar{S} = \beta = 2\sqrt{3}(b/L)$. If the span is slender, the depth-to-span ratio, b/L, is at most 1/10 down to 1/20, setting β in the range one-sixth to a third; $1 - \beta^2 \approx 0.88 \rightarrow 0.97$ and M, at worst, is 88% of M_P. We can adjust the safe limit of load accordingly, but the reduction is marginal provided we maintain a high slenderness – for any shape of cross-section.

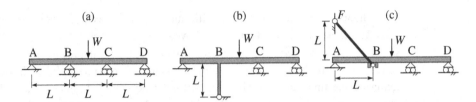

Figure 16.2 Three versions of a three-span long beam loaded centrally. (a) Conventional roller supports. (b) Leg support and full beam junction at B. (c) Tension tie at B which resists downwards pushing but not uplifting.

16.2   Support Variety

Figure 16.2 indicates three versions of a continuous beam with three equal spans, AB, BC and CD, each of length L and uniform cross-section. All main spans are identical, and each span BC is loaded vertically by a central point force, W.

The supports are all simple except for B, which becomes an upright column leg in Fig. 16.2(b) of length L but not of the same M_P. In Fig. 16.2(c), it is replaced by a tension tie, BF, inclined at $45°$ and of nominal length $\sqrt{2}L$. The lower end is nestled inside a pair of stopper blocks for location rather than being pinned through a hole in the main span directly above, which preserves a full cross-sectional plastic moment. This end can also slide downwards relative to the blocks if the beam uplifts locally to avoid compression and axial buckling; think of how the beam deflects if the direction of W is reversed.

The beam in Fig 16.2(a) is twofold redundant according to the bending moments at supports B and C. The free case replaces them by pinned connections, and the span BC cannot transmit moments to the others. The corresponding bending moment diagram in Fig. 16.3(a.i) has a negative peak of $-WL/4$.

Reinstating the moments in Fig. 16.3(a.ii) gives the separate reactant lines with positive salient values of M_1 and M_2 which can, of course, be combined into a single profile. The final bending moment diagram is shown in Fig. 16.3(b) with an extra negative salient value, M_3, equal to $-WL/4 + M_1/2 + M_2/2$.

Drawing each salient value to its respective plastic limit first gives $M_1 = M_2 = M_P$ and, consequently, $-WL/4 + 2M_P/2 \geq -M_P$, returning $W \leq 8M_P/L$.

This is the most straightforward of the three. The column in the second beam causes an extra, third redundancy, which we can think of as the top bending moment it locally applies to B in the horizontal span. At this member junction, there are thus three local bending moments, M_1, M_2 and M_3, Fig. 16.4(a): for their assumed directions, moment equilibrium of the junction sets $M_3 = M_2 - M_1$.

The reactant profile therefore has four salient values including M_4, Fig. 16.4(b). Note that these still define three redundancies because of the extra junction equation. The profile is drawn as usual on the tension side of the members, with the junction displaying the step change between M_2 and M_1; the free case is shown as inset. When combined in Fig. 16.4(c), the final M_5 value is equal to $-WL/4 + M_2/2 + M_4/2$.

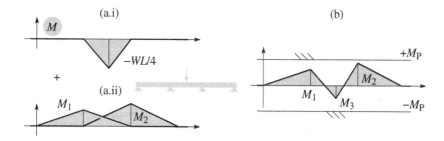

Figure 16.3 (a.i) Free bending moment profile for the beam in Fig. 16.2(a) after inserting pins at B and C; (a.ii) reactant profiles after reinstating bending moments at B and C. (b) Combined bending moments sitting between fully plastic limits.

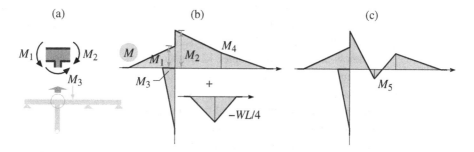

Figure 16.4 (a) Moment equilibrium performance of the junction at B in Fig. 16.2(b). (b) Corresponding reactant bending moment profiles and, inset, free profile from having pins at B and C. (c) Final bending moment profile.

Obviously, we can set $M_4 = +M_P$ and $M_5 \geq -M_P$, but not all of the moments at the junction can be set to their fully plastic values. For example, if M_3 is taken to be zero, $M_1 = M_2$ equals M_P, say, which gives the same result for W as for the first beam.

But what of the column response? Setting M_3 to be zero implies no moment transfer to the leg; it can be very thin indeed but not without buckling in the extreme because it is surely compressed axially. Notwithstanding, there will be some (elastic) moment transfer in practice as we build up to maximum load, and a thin column might be expected to yield first (in bending) at the top.

At this point in the discussion, it is useful to think about collapse mechanisms (but without calculating Upper Bounds) for a sense of how the moments perform. Also, if the column is indeed thin, its fully plastic capacity is going to be smaller than the other two, and we should set M_3 to a small fraction of M_P. Declaring $M_2 = M_P$ again obviates M_1 to be less than M_P from $M_3 = M_2 - M_1$.

Ideally, the plastic hinge forms on the M_2 side of B in span BC as shown in Fig. 16.5(a), giving a viable collapse mechanism and a load equivalent to the previous Lower Bound, $8M_P/L$. If, however, a plastic hinge forms instead in the column, Fig. 16.5(b), there is no viable mechanism because this hinge does not disrupt the

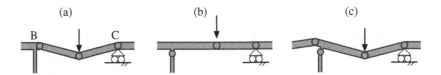

Figure 16.5 (a) Local collapse mechanism of the span BC in Fig. 16.2(b): note the left-side plastic hinge is just beyond the column leg. (b) Left-side hinge is now placed at the top of the column leg. (c) Collapse only proceeds, however, if another plastic hinge forms in the beam above, to the left of the leg.

continuity of the main span. A hinge must, therefore, form on the side of B in span AB, Fig. 16.5(c), with $M_1 = M_P$, for collapse to proceed.

However, M_2 being equal to $M_1 + M_3$, is now larger than M_P, which cannot be so: we must revert to the situation in Fig. 16.5(a) with the column remaining intact, no matter how thin, and M_2 governing the safe response entirely.

We know already from Chapter 12 that the column thinness precisely enables the beam moments to dominate equilibrium, and its low bending stiffness overall accords a rotational freedom similar to a simple support. But we emphasise again that the limit on thinness is dictated by the axial force capacity, either in compressive buckling or yielding, which much be larger than the actual column force.

The tension tie in Fig. 16.2(c) has been removed altogether and replaced by a pair of tensile force components, $T/\sqrt{2}$, in Fig. 16.6(a) which still enable point B on the main beam to move freely in the vertical (and horizontal) direction. There are now two redundancies again, the first being the bending moment at C and second the tension force, T (or equally, the bending moment beneath it at B).

The free span, AC, is now longer compared to the previous two cases, with a peak moment $M_1 = -3WL/8$ and reducing to $2M_1/3$ at point B, Fig. 16.6(b). The first reactant profile from T alone generates hogging moments with a peak value, $M_2 = \sqrt{2}TL/4$, and $M_2/2$ at the position of W, Fig. 16.6(c). The second reactant has M_3 at C and corresponding indicated values at B and W as shown, Fig. 16.6(d).

The final combined profile is drawn in Fig. 16.6(e) with salient values labelled; clearly M_3 should be set equal to M_P. Substituting this value into the other two salient values and defining dimensionless $t = \sqrt{2}TL/4M_P$ and $w = WL/8M_P$, we arrive at their inequality performances as:

$$-2w + t \leq 1/2, \quad -3w + t/2 \geq -7/4, \tag{16.3}$$

with the condition also that $t \geq 0$ to be effective in tension.

Taking $t = 1$ as an example, these insist $w \geq 1/4$ and $w \leq 3/4$, *i.e.* a range of solutions for W between $2M_P/L$ and $6M_P/L$. Otherwise, plotting the inequalities as lines in (t, w) space, it can be shown that they converge and simultaneously satisfy at $(t, w) = (5/2, 1)$. This coordinate declares $W = 8M_P/L$, our original best result for the first case, and the tension tie must have a large enough cross-sectional area to be able to furnish a limiting value of $5\sqrt{2}M_P/L$, where it now behaves like the original roller support.

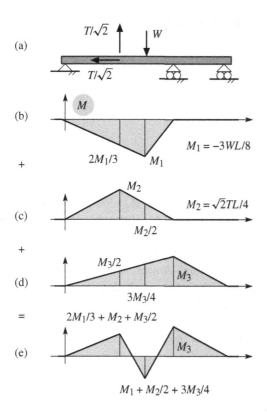

Figure 16.6 (a) Loaded beam from Fig. 16.2(c). (b) Free bending moment profile after removing T and the bending moment at the second support. (c) Moment profile after reinstating T. (d) Then, after reinstating moment at second support. (e) Final bending moment profile.

16.3 Final Remarks

An ordinary simple support is expected to furnish a 'rigid' vertical constraint whilst freely affording rotation. The second and third cases above, however, conflate these kinematic expectations: if the tension tie yields, the beam can move vertically at point B; if the column leg is not thin enough, moment transfer across this point is compromised.

As such, the ultimate performance of the main beam is reduced for contrasting structural expectations: the tie has to be thick enough to avoid yield, and the column should be thin. The latter increases the risk of axial buckling, but this has to lie outside the Lower Bound route, reinforcing again that we cannot discard elastic analysis altogether.

The horizontal tension component at B will also reduce the local plastic bending capacity, *c.f.* Chapter 15, requiring us to lower M_P. The same slenderness argument as per shear force previously applies in view of a marginal reduction; however, a slim span makes AB more prone to axial buckling under the compressive force $T/\sqrt{2}$, and we should consider both the reduced capacity and buckling in a trade-off analysis.

17 The Third Dimension

Our structures so far have been planar, deforming in their own plane. This does not restrict our view of practical three-dimensional structures: we can treat their planar walls or frames by our previous methods noting the importance of their *stability* when connected together; such analysis is beyond our scope for now.

We can, however, adopt a *pseudo* three-dimensional viewpoint by considering planar structures loaded *out of plane*. These are so-called *grillages*, made from interconnected beams and columns – and not from pin-jointed bars – because they resist loads by bending *and* by torsion (and by axial effects).

We draw them in isometric view for consistency, and we adopt Maxwell's right-hand screw rule for conveying the direction of moments and torques without ambiguity. For example, the planar right-angled cantilever in Fig. 17.1(a) is not a grillage when loaded in-plane by the tip force, F; but the directions of positive bending moment and shear force, M and F, at the corner junction are obvious from the usual planar considerations, Fig. 17.1(b). The cross-sectional proportions are assumed to be small compared to the overall size so that reactions between portions occur essentially at a junction *point*.

The same cantilever is now shown in isometric view in Fig. 17.1(c). The built-in base end, G, has been replaced by a horizontal support plate, able to resist any lateral movements or rotations of column bottom within its own plane. The column and beam elements are distinguished by labelling their ends or junctions as G, H and J, and the free-body diagram in Fig. 17.1(d) shows the same portions, GH and HJ, as in (b).

The bending moment, M, is highlighted two ways, firstly, by the usual planar curved arrows. Their directions, specifically the plane in which they curve, GHJ, is not easily conveyed without annotation, which is inefficient. Alternatively, they are drawn as vectors normal to the plane, pointing in the direction of the right-hand thumb when adjacent fingers curve in the moment direction. They are also drawn double headed, in order to distinguish from forces, but like forces their vectors can be added for moment equilibrium (c.f. bi-axial bending in Chapters 14 and 15).

To highlight as such, the tip force is replaced by a torque, C, applied parallel to HJ, Fig. 17.1(e). Beam HJ is equilibrated by an equal and opposite torque at H, which is reacted onto column GH at both its tip and base, and all four components are parallel and planar. Column GH is effectively loaded by equal and opposite end moments,

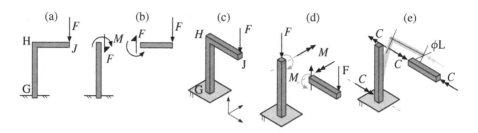

Figure 17.1 (a) Right-angled cantilever loaded by a planar tip force. (b) Corresponding stress resultants, represented in planar form. (c) Isometric view of the same cantilever: roman (upright) labels denote specific points. (d) Corresponding stress resultants with moments and torques as double-headed vectors. (e) The same cantilever tip-loaded by an axial torque, which deforms out of plane.

causing it to deflect laterally, as shown; beam HJ remains straight but twists, where the twist rate, ϕ, is equal to C/GJ from Chapter 14.

Recall that G is the shear modulus and J the polar moment of inertia, and together, GJ is the torsional stiffness, c.f. EI for bending stiffness. For uniform torque along the beam, the relative rotation between the ends equals the product of twist rate and beam length: we shall use this in the following examples.

17.1   In Series or Parallel?

The horizontal right-angled cantilever, GHJ, carries a vertical tip force, F, at J in Fig. 17.2(a). Both beam portions GH and HJ have equal lengths L, the same uniform cross-section and are made from the same material. We wish to calculate the vertical deflection of the tip, δ_J, and compare its stiffness to an ordinary cantilever of the same proportions and length L.

Each portion is shown as a free body in equilibrium in Fig. 17.2(b), where there are equal and opposite vertical shear forces at the junction, H, leading to the same at base G. There is a bending moment, C, about the base of HJ within the vertical plane, where a side-view would confirm $C = FL$. It points away from the base in order to resist the tendency of HJ rotating downwards under F, and is reacted onto GH by an equal and opposite torque, twisting the beam in the same direction. There is a final torque reaction of C at the base.

The displacement of J is divided into three vertical components in Fig. 17.2(c). The first, δ_1, is due to GH alone deflecting as a tip-loaded cantilever, with the original axis of HJ translating linearly in the same plane. The end of GH also rotates under twisting from C, to the give the leveraged displacement δ_2 equal to the axial tip rotation of GH, shown as θ, multiplied by L. The third component is the elastic cantilever deflection of HJ relative to GH, as if GH were rigid and did not deform. Their final expressions can be verified to be:

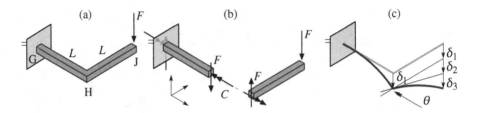

Figure 17.2 (a) Right-angled cantilever grillage, tip-loaded out-of-plane. (b) Equilibrium performance of straight free-body sections. (c) Displacement components from separate deformation actions.

$$\delta_1 = \frac{WL^3}{3EI}, \quad \delta_2 = \underbrace{\frac{C}{GJ}L}_{\text{GH rotation}} \times L, \quad \delta_3 = \frac{WL^3}{3EI}$$

$$\rightarrow \quad \delta_J = \delta_1 + \delta_2 = \delta_3 = \frac{WL^3}{3EI} \cdot \left[1 + \frac{3}{\beta} + 1\right]. \tag{17.1}$$

The pre-factor above is the tip deflection of an identical cantilever of length L. The unity terms inside brackets are, therefore, the bending contributions from GH and HJ, and the middle term measures the *relative* twisting contribution using the stiffness ratio, GJ/EI, equal to β.

Typical values of β are given later, but we see that δ_J can become very large if β is very small, and infinite when zero; GH offers no torsional resistance and we have a mechanism. Conversely, increasing β reduces the displacement down to the case when β is infinite, with only bending components, which are finite nonetheless.

The stiffness in bending *and* in torsion are, therefore, both important in determining the overall displacement. In fact, we might say there is a 'series'-like performance compared to, say, simpler discrete spring systems: the least stiff element contributes the largest displacement and thus governs overall.

17.2   Much Torquing

The previous grillage is now modified by building in the free end, see Fig. 17.3(a), which in turn makes it statically indeterminate – from constraining further an already determinate structure, and thus more challenging to solve. It is loaded by a central point force, F, at H, and we wish to find the displacement, δ_H.

The grillage is first divided into two cantilever halves about H, Fig. 17.3(b). Vertical force equilibrium of an infinitesimal central element indicates a pair of shear forces, which must be equal from the symmetrical nature. The tip displacement of each half for this loading alone is a standard result, which provides a provisional estimate: $\delta_H = (F/2)L^3/3EI$.

The tip of each beam also rotates within its own vertical plane, which must exactly match the twisting rotation of the other tip in the same plane. This is wrought by a

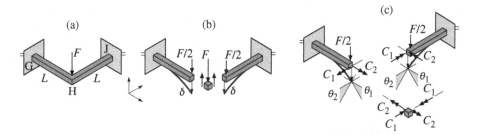

Figure 17.3 (a) Indeterminate cantilever grillage loaded centrally. (b) Free-body shear force loadings for each cantilever leg and central element. (c) Additional moments and torques applied to each leg and to central element.

pair of torques orthogonal to each other from the right-angled layout, reacted in turn by bending moments at each tip, see Fig. 17.3(c). The torque directions are assigned positively in the direction of the twisting rotations, denoted as θ_1 and θ_2, giving moments which resist the end rotations of their respective beams.

Moment equilibrium of an infinitesimal element at H, Fig. 17.3(c), only confirms that each torque-moment pair of components, C_1 or C_2, is equal but not to the other pair. However, from the symmetry of layout and deformation, we can say that $C_1 = C_2$ ($= C$, say) and $\theta_1 = \theta_2$ ($= \theta$): this will not be so for unequal cantilever lengths or when each half differs in EI (or GJ). Standard cantilever bending results from Chapter 10 give:

$$\theta = \frac{(F/2)L^2}{2EI} - \frac{CL}{EI}, \quad \delta_H = \frac{(F/2)L^3}{3EI} - \frac{CL^2}{2EI}. \tag{17.2}$$

The torque tip rotation, CL/GJ, equals θ, allowing us to find C in terms of F, i.e. $FL^2/4EI - CL/EI = CL/GJ$, setting $C = \beta FL/4(1+\beta)$ with β our stiffness ratio from earlier. Substituting C back into the expression for δ_H and re-arranging:

$$\delta_H = \frac{FL^3}{6EI} \cdot \left[1 - \frac{3\beta}{4(1+\beta)}\right]. \tag{17.3}$$

The pre-factor, $FL^3/6EI$, arises when β is zero: it therefore pertains to our provisional result, which neglects any torsional connection – as if the two halves were connected by a 'ball-joint' transmitting forces but not moments or torques in three dimensions. When the halves are connected as they are, β cannot be zero and its value depends on each of G, E, I and J; imagine then that β *can* vary in value in order to assess the effect of the torsional stiffness on displacement.

For very large values of β, the term $1 + \beta \approx \beta$ and the brackets value is approximately equal to 1/4. For a torsionally rigid grillage, the displacement is around 1/4 of the case when $\beta = 0$: despite the rotations θ_1 and θ_2 being inhibited, each half is still able to deflect. The range of δ_H, therefore, lies between extrema of cantilever bending alone where the tip is free to rotate or not at all.

In practical thin-walled sections, J is typically equal to $2I$: when made of isotropic material, where $G = E/2(1+\nu)$, then $\beta \approx 1/(1+\nu)$. For most metals, ν is around 1/3, giving $\beta \approx 3/4$ and setting the brackets value in Eq. (17.3) to be around 2/3; *c.f* our extreme case values of 1/4 and unity. These are significantly different displacement performances, underpinning again the role that torsion plays in such problems.

17.3   Limited Torquing

The horizontal T-section, GHJK, in Fig. 17.4(a) is loaded vertically by a central point force, F, at H. This junction connects three identical cantilevers of length L, two of which are shown as free bodies in Fig. 17.4(b): we do not need to show the extra 'leg', HK, in this view because its response is symmetrical to that of HJ.

The shear forces in local elemental equilibrium with F are symmetrical only with respect to beams HJ and HK. These are denoted as R and are different from that on the side of GH, equal to P: thus $F = 2R + P$. Symmetry also precludes twisting of GH as well as any rotation at H along HJ, but the tip of GH can rotate in its own vertical plane, where twisting of the entire beam, JK, must follow from compatibility.

Because twisting inhibits the free tip rotation of beam GH, a resistive bending moment, C, is applied to H. This is equilibrated by a moment pair, $C/2$, applied to the central element of section, which are reacted as torques at H on both ends of HJ and HK. Orthogonal to the same ends, we must also include bending moments, M, which oppose the rotations induced by each of R, to ultimately ensure zero gradient at H along JK.

The torsional rotation at H for HK and HJ (as shown) is θ_1 equal to $(C/2)L/GJ$. This is also the tip rotation of GH bending as a cantilever under P and moment C; the corresponding displacement is δ_H, whence:

$$\theta_1 = \frac{PL^2}{2EI} - \frac{CL}{EI}, \quad \delta_H = \frac{PL^3}{3EI} - \frac{CL^2}{2EI}. \tag{17.4}$$

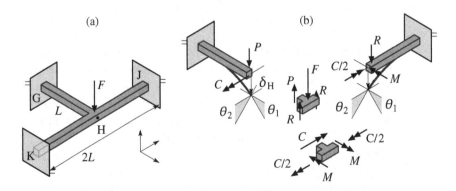

Figure 17.4 (a) Indeterminate T-section grillage loaded centrally. (b) Equilibrium and kinematical performances of constituent cantilever legs and central element.

From equating the θ_1 expressions, we find $PL^2/2EI = CL/EI + CL/2GJ$. Replacing GJ using βEI from before, then re-arranging in terms of C and simplifying, it can be shown that $C = PL\beta/(1 + 2\beta)$. This gives the displacement in terms of P as

$$\delta_H = \frac{PL^3}{3EI} - \frac{PL\beta}{1+2\beta}\frac{L^2}{2EI} = \frac{PL^3}{EI} \cdot \left[\frac{2+\beta}{6+12\beta} \right]. \tag{17.5}$$

The rotation of the tip of beam HJ within its own plane is θ_2 even though we know it to be zero from symmetry. Both M and R contribute to this and to the displacement, δ_H, as:

$$\theta_2 = \frac{RL^2}{2EI} - \frac{ML}{EI}, \quad \delta_H = \frac{RL^3}{3EI} - \frac{ML^2}{2EI}. \tag{17.6}$$

The first of these sets $R = 2M/L$ when $\theta_2 = 0$, and thus $\delta_H = RL^3/12EI$ after substituting into the second equation. There are now two expressions for δ_H in terms of P or R, which, after equating, reveal: $P = (1 + 2\beta)R/(4 + 2\beta)$.

We can now substitute P from above into $F = 2R + P$ to express F purely in terms of R and thence δ_H. Our final result is expressed in the same product format as before, by a pre-factor due to bending only without torsional coupling and the sum of unity and a nonlinear β term:

$$\delta_H = \frac{FL^3}{27EI} \cdot \left[1 - \frac{9\beta}{54 + 36\beta} \right]. \tag{17.7}$$

When β is very large, the reduction in displacement via the brackets value is 9/36, *i.e.* 25 %; *c.f.* 3/4 for the previous example. The role of torsion this time has clearly reduced, evinced in part by beam GH not twisting from the outset. Even though HK and HJ do twist, each must also bend but comply with zero gradient at H, which stiffens the bending response in comparison.

What is more striking, however, is that for the typical thin-walled value of β equal to 3/4, the displacement only reduces only by around 8% compared to the torsion-free case. Ignoring torsion altogether does not take us that far from the exact solution, which makes for a simpler calculation; but we may only conclude this having performed the exact calculation.

17.4   Final Remarks

Grillages belong in virtually every practical structure. We have simplified their loading to be at member junctions, to exploit standard cases along with symmetry and obvious compatibility conditions. Bending and torsion are clearly coupled, and the overall stiffness is a composite measure usually combining elements in 'parallel' (*c.f.* the series performance of the first right-angled cantilever): if one stiffness measure becomes severely diminished, there is no excessive compliance in the rest of the structure. The ratio of bending-to-torsional stiffness is thus a key relative parameter for grillage performance.

Part VI

Deliberately Deformed

18 How Many Collapse Mechanisms?

We saw in Chapter 6 how a Lower Bound approach guarantees a safely loaded truss for any equilibrium solution not violating the yield condition; in Chapter 16 we studied the same safe loading of beams and frames. For all structures, partial yielding is permissible in the sense of broaching the yield-free limit, but not critically throughout.

On the other hand, a structure collapses because enough of the structure has now yielded, usually from unexpected loading levels. In discrete beams and frames, collapse manifests as mechanistic motion; of largely rigid, relatively undeformed body parts, interconnected by localised highly-plasticised regions. Understanding how collapse proceeds is important; or we can accept that the structure is already in this state in order to make sense of its *ultimate* capacity.

Other assumptions are necessary for tractable insight. The deforming geometry is simplified by first treating the focussed plastic deformation as *point hinges* in slender members. Any elastic deformation is neglected in comparison to the relative plastic rotation between members, which accords a constant specific (per unit radian) resistance denoted by M_P, the fully plastic moment, *c.f.* Chapter 15.

These rotations are large enough for a distinct collapsed shape but, equally, are assumed to be relatively small for simplifying calculations. An important outcome is collapse load expressions devoid of kinematical parameters, *i.e.* fixed loads that depend on the shape *viz.* distribution of collapse mode but not its amplitude.

Loading proceeds in a ramp-like fashion from zero until the first collapse mechanism emerges. No unloading before collapse occurs is allowed in case of yielding somewhere at an interim peak load without evidently collapsing; any re-loading from zero may change the complexion of collapse and, as such, failure is not *incrementally* achieved but rather in a single *static* step.

The structure's collapse mechanism always behaves as a single degree-of-freedom system with a minimum number of plastic hinges. More hinges accord more freedoms and more complex motions that, however, do not have to be assessed because the Upper Bound Theorem favours the simplest collapse mode.

To determine their number, we note that there is no difference between a plastic hinge and a frictionless pin, kinematically. One approach in Chapter 8 on finding the number of redundancies inserts ever more pins into an indeterminate frame until it just becomes a mechanism. Equally, the minimum number of plastic hinges is the same, being one more than the number of redundancies. In this sense, the overall structure collapses *regularly*, but where *partial* collapse is feasible, fewer hinges are needed.

The hinges must also be located around the structure, remembering they form because of excessive bending. The elastic bending moment profile provides a useful starting point because its ordinate values 'stretch' with the load increasing until the largest bending moment reaches and saturates at M_P. The bending moments continue to rise elsewhere, as such redistributing (*c.f.* Chapter 6) the spare elastic capacity to mete increasing load. The detailed sequence is not important for it requires careful calculation; we shall be satisfied instead with salient elastic values as prime candidates for hinge locations.

The junctions between members or support points are where peak bending moments typically occur. The number of viable hinge locations therefore increases with the level of structural detail, leading to multiple hinge combinations whatever the minimum number of hinges required. Postulating several collapse mechanisms is usual in order to calculate a range of collapse loads and combinations.

This calculation balances the external work effort with internal dissipation. For the correct collapse mechanism, the structure is just on the brink of failure because statical equilibrium is also satisfied: as a corollary, the margin of safe loading is now zero even though the structure still stands. Without knowing about equilibrium – not obligated by the Upper Bound Theorem – we cannot pronounce ultimately upon the safety of the structure; but within our fairly simple problems, we can examine all possible collapse mechanisms including the correct one.

We now step through this process by example. Understanding how many plastic hinges are needed and where they should occur is considered first before developing a systematic method for computing the motions within a collapsing frame – an exercise that can often frustrate. We also deal with how to present findings compactly for multiple collapse modes.

18.1   Plastic Candidacy

The two-span beam in Fig. 18.1(a) is indeterminate with one redundancy. A single force is applied to the first span; precisely where is not important except that it generates a negative peak moment, M_1, at the same position. The reaction forces guarantee a piece-wise linear profile throughout, with a positive peak value M_2 over the internal support and no bending moment at both ends.

When the applied force increases, the profile stretches transversely until M_1 and M_2 reach their fully plastic limits, giving two plastic hinges – one more than the single redundancy for the beam. The order in which they attain a value of M_P is not important: after the first hinge forms at the internal support or underneath the load, the other peak elastic value will be first to reach the plastic limit.

The signs of our peak bending moments obviously match the deformation across each hinge. Underneath the applied force there is sagging (negative) behaviour, and over the support there is hogging. If in doubt about the moment signs, we can sketch provisional elastic deflections for a sense of how the direction of bending proceeds along the beam, as per Chapter 7.

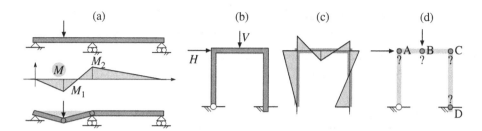

Figure 18.1 (a) Two-span beam, corresponding general bending moment profile and collapse mechanism with two plastic hinges. (b) Portal frame loaded horizontally and vertically. (c) Corresponding bending moment profile. (d) Salient values indicate prime locations of plastic hinges during collapse.

A more intriguing example is the doubly-loaded square portal frame in Fig. 18.1(b), which has a single pinned foot. The other foot is built in, and overall the frame is doubly redundant. Forces H and V are applied as shown, giving piece-wise linear bending moments drawn as positive values on the tensile side of the frame, Fig. 18.1(c).

There are four salient values and, thus, four likely locations for M_P, labelled A, B, C and D; in each upper corner (one of which coincides with H), underneath V and at the right-side base, see Fig. 18.1(d).

Given two redundancies, only three hinges within the four locations are needed for collapse. Four candidate mechanisms therefore arise when we count all possible combinations, as shown in Table 18.1.

The first of these is our *beam* mechanism with three hinges at A, B and C in the upper beam alone, Fig. 18.2(a); only V performs any useful work. Strictly, these hinges can be located on the column side of the corners adjacent to them because of the point-hinge assumption. Nonetheless, the columns remain undisturbed and collapse is partial even though the maximum number of hinges is needed.

Beam collapse occurs when V is much larger than H. If reversed, the frame will tend to move horizontally and thus *sway* as a four-bar mechanism under H alone. The corresponding hinge positions, A, C and D, are shown in Fig. 18.2(b) with every member moving in regular collapse.

When the forces are roughly equal, their effects are not easily separated, leading to a collapse mode *combining* the sway and beam modes. The hinge positions in Fig. 18.2(c) are at B, C and D, and the corner at A does not deform. Clearly H and V both do useful work, calculated momentarily.

The final mechanism relocates the hinge at C to A, Fig. 18.2(d), to give another combined mechanism. The right-angled portion to the right of V now rotates clockwise as a rigid body, causing point B to move upwards. Negative work is thus performed by V, which does not favour or promote collapse compared to the other modes. This collapse mode can be discounted, and the remaining three modes of *beam*, *sway* and *combined* collapse are usually sufficient for answering most problems.

Table 18.1 Plastic hinge positions in Fig. 18.2.

A	B	C	D	name/Fig. 18.2
X	X	X	-	beam/(a)
X	-	X	X	sway/(b)
-	X	X	X	combined/(c)
X	X	-	X	combined/(d)

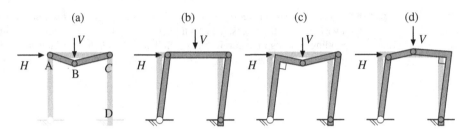

Figure 18.2 (a) Beam collapse mechanism. (b) Sway mechanism. (c, d) Combined beam and sway collapse mechanism.

18.2   Moving Hinges

The correct collapse mode equates to the smallest collapse load. When there are several modes to choose from, the external work/internal energy calculation is performed each time for static hinge positions. Sometimes the hinge position varies because the loads vary in position, where the collapse outcome is now a function of both.

Alternatively, we may have to locate a hinge by a beam-wise coordinate because we do not know the bending moment variation precisely – because an exact equilibrium solution is not necessary; simple calculus then affords the exact position from the minimum collapse load (and thence the correct equilibrium distribution).

For example, a horizontal force is applied to a square portal frame at a vertical variable distance, x, above ground in Fig. 18.3(a). The frame has two built-in feet and, thus, three redundancies, giving a mechanism requirement of four plastic hinges. These are located in the sway mechanism in Fig. 18.3(b) at the feet and corners, of small relative rotations θ across them.

The point of application of H displaces δ equal to $x \sin \theta$, which is approximately $x\theta$, giving an amount of external work of $H \cdot x\theta$. Since the plastic resistance from M_P is a fixed amount per unit rotation, the internal dissipation simply multiples each fully plastic value by the relative rotation across that hinge; here, it is $4 \cdot M_P \cdot \theta$ for identical frame members. Equating both efforts and eliminating a common θ between them, we find our first critical load, $H = H_1$ equal to $4M_P/x$.

Another feasible sway mechanism sees the two upper hinges forming in line with H, Fig. 18.3(c), which is a likely position of maximum bending moment; the work-energy calculation, however, yields the same critical load as H_1, and either sway mechanism is equally likely.

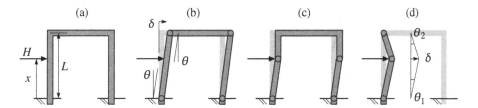

Figure 18.3 (a) Side-loaded square portal frame. (b) Sway collapse mechanism with upper plastic hinges in the corners. (c) Another sway mechanism with upper hinges on the line of action of loading. (d) Local beam mechanism column-side.

The formation of a corner hinge *and* one next to H comes from the beam mechanism in Fig. 18.3(d). The enunciation is the same despite forming in the column, and only three out of four hinges are needed. The small, relative rotations in the bottom and top hinges are θ_1 and θ_2, respectively, where the simple triangular geometry shown gives δ equal to $x\theta_1$ or to $(L-x)\theta_2$, which sets the ratio of rotations.

The external work done is again $H \cdot x\theta$, and the relative rotation across the internal hinge is the *sum* of corner rotations, $\theta_1 + \theta_2$, which can be confirmed by excising the originally straight column and performing these rotations in turn about the internal hinge. The internal energy dissipated is thus $M_P[\theta_1 + (\theta_1 + \theta_2) + \theta_2]$.

Substituting for θ_2 from the ratio of rotations, we find $\theta_2 = \theta_1 x/(L-x)$; balancing the external work and dissipated energy, we eliminate θ_1 and arrive at a new expression for H, now equal to H_2. Making dimensionless by multiplying H by L/M_P and denoting as $\bar{H}$, whereupon:

$$Hx = M_P\left[2 + \frac{2x}{L-x}\right] \quad \rightarrow \quad \bar{H}_2 = \frac{H_2 L}{M_P} = \frac{2}{\lambda}\left[1 + \frac{\lambda}{1-\lambda}\right]. \tag{18.1}$$

Here $x/L = \lambda$, which lies in the range zero to unity; for the original sway, we can write, $\bar{H}_1 = 4/\lambda$.

The variations of $\bar{H}_1$ and $\bar{H}_2$ with λ and thence their *interaction* are plotted in Fig. 18.4. The curves coincide at the halfway point, $\lambda = 1/2$, where briefly either collapse can occur, but not together. For λ smaller, the curve for $\bar{H}_2$ lies below $\bar{H}_1$ and beam-collapse prevails; when H is applied higher than halfway, swaying via $\bar{H}_1$ takes precedence.

In the second varying hinge case, a propped cantilever collapses in Fig. 18.5(a) under a uniformly distributed loading of W. From Chapter 7, the bending moment varies quadratically in general, from a positive peak value at the base to a negative minimum near the middle (but not exactly so), and to zero at the tip support. The single redundancy, again, demands two plastic hinges at most.

To complement the hogging hinge at the base, a second hinge lies a distance x along, giving unequal collapsing parts. The rotations of the beam ends are compatible with the transverse displacement, δ, of the second hinge from $\delta = x\theta_1 = (L-x)\theta_2$.

Evaluation of the internal dissipation is straightforward, $M_P[\theta_1 + (\theta_1 + \theta_2)]$; for the external work, we divide the effort between the rotating parts. Over the first part,

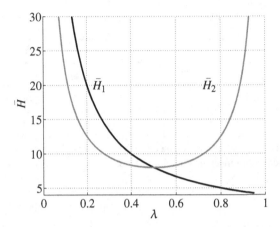

Figure 18.4 Variation of dimensionless collapse loads, $\bar{H}_1$ and $\bar{H}_2$, with height position λ ($= x/L$) of force applied to the portal frame in Fig. 18.3.

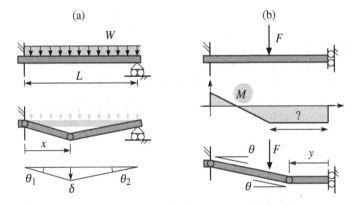

Figure 18.5 (a) Collapse mechanism and kinematics for uniformly loaded cantilever. (b) The same for another indeterminate cantilever, with restrained tip rotations.

the total load is $(W/L)x$, and its centroid halfway along at $x/2$ displaces vertically by $\delta/2$, using similar triangles within the displaced profile. For the second part, the load is $(W/L) \cdot (L - x)$ whose centroid also moves by $\delta/2$.

Multiplying each force and displacement and adding together returns the expression $[(W/L)x + (W/L) \cdot (L - x)] \cdot (\delta/2)$, simplifying to $W\delta/2$, which is precisely the average displacement across the profile, times W. As a result:

$$W\delta/2 = M_P[2\theta_1 + \theta_2] = M_P\left[\frac{2\delta}{x} + \frac{\delta}{L - x}\right]$$

$$\rightarrow \quad \frac{WL}{M_P} = \left[\frac{4}{\lambda} + \frac{2}{1 - \lambda}\right] \tag{18.2}$$

where again $\lambda = x/L$.

Minimising the right-side bracket by differentiating with respect to x and setting equal to zero sets the position of the hinge. This obviates $-2/\lambda^2 + 1/(1-\lambda)^2 = 0$ with soluble roots, $\lambda = 2 \pm \sqrt{2}$. Selecting the lower root, the hinge is located at $(2 - \sqrt{2})L \approx 0.58L$ inside the beam ($\lambda \leq 1$). The corresponding collapse load is $WL/M_P = 2\sqrt{3}/(3\sqrt{2} - 4) \approx 11.66$.

If we *had* guessed instead the halfway point for the second hinge (and thus reneged on its variable position via x), our estimated (dimensionless) collapse load would be $4/0.5 + 2/(1 - 0.5) = 12$, a difference of around 3% compared to the exact value for much less working.

Another singly redundant cantilever is given in Fig. 18.5(b). The tip is supported differently, by a vertical roller which suppresses end rotations by applying a reaction moment; transverse displacements are otherwise free.

The centrally applied point force produces a linear bending moment profile by itself, with a peak positive moment at the base tapering down to zero underneath F. The tip moment applies a uniformly negative bending moment throughout, and their superposition leads to the overall profile in Fig. 18.5(b). Over the second half of the beam, there is constant negative moment.

The first hinge is located at the base and acts in hogging. The lack of a singular negative value beyond F denies a specific location for the second hinge; it can lie anywhere to the right of F in theory. In fact, its general position from a coordinate y does not enter the work balance, unlike the previous example: F displaces $(L/2)\theta$ and the hinges dissipate $M_P[\theta + \theta]$ for relative rotations of θ. As a result, $F = 4M_P/L$.

In practice, the hinge is likely to form underneath F at the transition point from linear to uniform, and furthest away from the rotational constraint of the tip roller; this would make for an interesting experiment.

18.3 Instantaneous Centre

We return to the combined collapse mechanism of Fig. 18.2(c), re-drawn in Fig. 18.6(a.i). Because of its name, many superpose the separate kinematics of its constituent beam and sway mechanisms to find its motion. There is nothing wrong in doing this, but we must remember to cancel rotations in the top-left corner, which remains rigid and undeformed.

Our more general method starts by identifying the parts of the frame between hinges as separate bodies, which are drawn as general egg-shapes in Fig. 18.6(a.ii). Points of connection and rotation are identified and labelled A ... D, and coincide with those of the original frame.

Each egg body is now isolated in Fig. 18.6(b) and their prospective rotations are indicated as θ_1, θ_2 and θ_3: we also have to identify their *centres* of rotation. These are obvious when there is a fixed plastic hinge or pinned support, as per the grounded points A and D. The displacement of the connection points B and C in their respective bodies are found by multiplying their lever-arm lengths by rotation. For B we have $\delta_1 = \text{AB} \cdot \theta_1$, and for C, $\delta_3 = \text{DC} \cdot \theta_3$, both in perpendicular directions as shown.

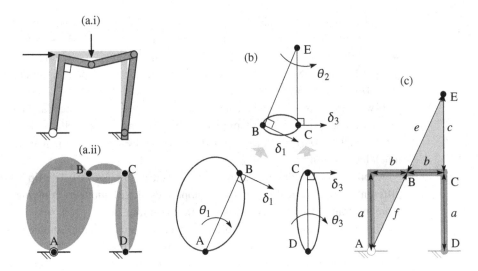

Figure 18.6 (a.i) Combined collapse mechanism for portal frame; (a.ii) identification of separate rigid bodies (egg-shaped) connected by plastic hinges and pin-joints (black dots) in the original configuration. (b) Key displacements, rigid-body rotations and locations of instantaneous centres. (c) Relevant original geometry for calculating kinematics.

The intermediate body appears to float between these with no obvious physical centre of rotation. But we can infer a virtual or *instantaneous* one from its displacements at B and C, which are identical to the other B and C points. Immediately we can append δ_1 and δ_3 as vectors to the present body.

Projecting hypothetical lever arms from B and C at right angles to the displacements, we have our rotation centre at their intersection, shown as E in Fig. 18.6(b). The direction of either displacement relative to E tells that the rotation of this body, θ_2, is anti-clockwise and opposite to the other two.

Since we have a single degree-of-freedom, we can declare one rotation to be independent, say θ_1, from which the other two can be calculated in relative terms. From the common displacements:

$$\delta_1 = AB \cdot \theta_1 = BE \cdot \theta_2 \quad \rightarrow \quad \theta_2 = \frac{AB}{BE} \cdot \theta_1 \text{ (a)},$$

$$\delta_2 = DC \cdot \theta_3 = CE \cdot \theta_2 \quad \rightarrow \quad \theta_3 = \frac{CE}{DC} \cdot \theta_2 = \frac{CE}{DC}\frac{AB}{BE} \cdot \theta_1 \text{ (b)}. \qquad (18.3)$$

Returning to our original frame layout in Fig. 18.6(c), we can find E directly having stipulated our plastic hinge positions and identified the various rigid bodies. We draw a line from A to B and project it beyond B: likewise, from D to C and beyond, until it intersects the first line at E.

The highlighted similar triangles provide another aid to calculating the relative geometric proportions. For example, we had $\theta_2 = (AB/BE) \cdot \theta_1$, where AB/BE is now f/e in the new sub-figure. But rather than compute the ratio of diagonal lengths, we observe that $f : e = a : c \, (= b : b = 1 : 1)$ and we can replace f/e by these ratios if easier to measure, infer or calculate.

There is one last calculation, or rather summation, to perform. The relative rotation at B is θ_1 plus θ_2 for their positive directions. Again, imagine the bodies either side as free-standing and undergoing these small rotations, and tracking how the rotation across B changes. For C, we have $\theta_2 + \theta_3$; and at D, there is only θ_3. In general, the rotations are simply added together but this simple thought experiment gives assurance.

18.4 Collapsing Frames

The collapse loads of the unequal frame in Fig. 18.7(a) are now calculated using the above scheme. It has a width L and unequal leg lengths, L and $3L/2$. A horizontal force H acts at the left corner and a vertical force V halfway along the middle beam. Three redundancies beg four plastic hinges for regular collapse, and salient positions of elastic bending moment are at points A, B, C, D and E.

The column rotations of the sway mechanism about each foot in Fig. 18.7(b) are different and equal to θ_1 and θ_2. The top beam moves horizontally, making the top hinge rotations the same as the base in each column; the energy dissipated is, thus, $M_P[\theta_1 + \theta_1 + \theta_2 + \theta_2]$. At H, the lateral displacement of the beam end is $L\theta_1$, which is matched by other end, $(3L/2)\theta_2$, giving $\theta_2 = 2\theta_1/3$. Equating the external work done and internal dissipation, we find $H = H_1 = 10M_P/3L$.

The beam mechanism involving V has three hinges at B, C and D but is not shown; we find $V = V_2 = 8M_P/L$. The combined mode is more challenging, Fig. 18.7(c), with hinges at A, C, D and E, assuming the upper left corner remains rigid. The frame sub-divides into three rigid portions, ABC, CD and DE, and the instantaneous centre of body CD comes from projecting the line AC to where it meets the projection of DE, Fig. 18.7(d).

From the displacement of C, $\theta_1 = \theta_2$, and from that of point D, $L\theta_2$ or $(3L/2)\theta_3$, giving $\theta_3 = 2\theta_2/3$. The horizontal effort is $H \cdot L\theta_1$, and the corresponding vertical displacement of V is the product of θ_1 and the perpendicular distance from A to its line of action, $L/2$. We can now assemble the work equation, denoting our third mechanism by subscript '3':

$$H_3 L\theta_1 + V_3(L\theta_1/2) = M_P \left[\underbrace{\theta_1}_{A} + \underbrace{\theta_1 + \theta_2}_{B} + \underbrace{\theta_2 + \theta_3}_{D} + \underbrace{\theta_3}_{E} \right]$$

$$\rightarrow \quad \frac{H_3 L}{M_P} + \frac{V_3 L}{2M_P} = \frac{16}{3}. \tag{18.4}$$

There are no other worthy mechanisms to explore, and the collapse load interaction is plotted in Fig. 18.7(e) with dimensionless (positive) force axes $HL/M_P \, (= \bar{H})$ and $VL/M_P \, (= \bar{V})$. Each collapse load equation is linear, and plotted they enclose a polygonal region highlighted in grey. Beyond this region, there is certain collapse for

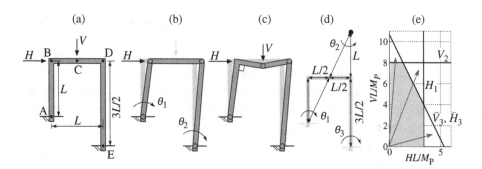

Figure 18.7 (a) Loaded stepped portal frame. (b) Sway collapse mechanism. (c) Combined mechanism. (d) Kinematics and geometry for the combined mechanism. (e) Interaction diagram of dimensionless collapse loads.

whatever pairing of H and V. Inside this region, the structure is safe – every viable collapse mechanism has been found.

Increasing the loading from zero when H and V are in a fixed ratio is tantamount to adding a line to Fig. 18.7(e) through the origin. Where it first meets a given collapse boundary defines the prescient mode for that ratio. These lines are appended as grey arrows: for shallow inclinations where $H \gg V$, the sway mechanism is favoured ($\bar{H}_1$), followed by the combined mode ($\bar{V}_3, \bar{H}_3$) and then the beam when $V \gg H$ ($\bar{V}_2$).

A second and final portal frame in Fig. 18.8(a) carries a vertical load of $2W$ uniformly distributed across its pitched roof, akin to its self-weight or to loading from, say, snow. The pitch height is h and width is $2L$, and the sub-frame columns are L tall. One foot is pinned, giving two redundancies and three plastic hinges for regular collapse.

The obvious symmetry of loading suggests the symmetrical collapse mode shown in Fig. 18.8(b.i). The four plastic hinges required are one more than is necessary, but symmetry ensures a single degree-of-freedom: the columns splay equally outwards as the roof pitch flattens.

Note that any previous overturning of columns was accomplished only by having external sideways loading. Here, it stems directly from the geometry of deformation: as each half of the pitched roof rotates, its projected length onto the horizontal increases in proportion to the rotation, forcing a compatible transverse displacement of each column head.

There is also an asymmetrical splaying mode, Fig. 18.8(b.ii), where only three hinges are required and the left-side column remains intact. The right-side column collapses in preference to the left, because of its pinned foot: the left, moving instead, demands an extra plastic hinge at its built-in foot and thus extra work during collapse (which raises the collapse load).

The comparison of these two modes points to the role of symmetry during collapse – or rather the asymmetrical preference, as we find out. Four separate rigid bodies are defined by the junction points A ... D in Fig. 18.8(c), with symmetric rotations about the centre-line.

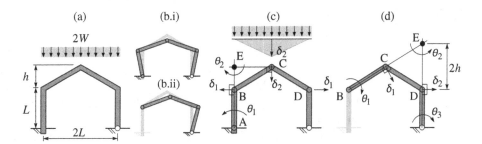

Figure 18.8 (a) Uniformly loaded, pitched portal frame with a pinned right foot.
(b.i) Symmetrical and, (b.ii), asymmetrical collapse mechanisms. (c) Kinematics and geometry
for (b.i). (d) Similarly for (b.ii).

The rotation outwards of AB is θ_1, and body BC rotates by θ_2 about its instantaneous centre E. The latter is found in the usual way, by projecting lever arms normal to the displacements at B and C, assuming that C moves purely vertically. Their intersection lies directly above B, and from its displacement δ_1 we find $L\theta_1 = h\theta_2$.

The external work performed by $2W$ considers the effort of each of its halves, W. Its displacement field is linear, Fig. 18.8(c), with a central peak, δ_2, equal to $L\theta_2$. Each half-load centroid therefore moves $\delta_2/2$, giving a total effort, $2 \times W\delta_2/2$. The overall work-energy balance now reads:

$$W\delta_2 = M_P \left[\underbrace{\theta_1}_{A} + \underbrace{2(\theta_1 + \theta_2)}_{B, D} + \underbrace{2\theta_2}_{C} \right] \quad \rightarrow \quad \frac{WL}{M_P} = \frac{3h}{L} + 4. \qquad (18.5)$$

Three rigid bodies move in asymmetrical collapse in Fig. 18.8(d). Body BC rotates downwards by θ_1, and CD by θ_2 about its instantaneous centre E, constructed as shown. The compatible displacement of C between them sets $\theta_1 = \theta_2$. Point D moves horizontally by $L\theta_3$ relative to the pinned foot but also by $2h\theta_2$ relative to E, whence θ_3 in terms of θ_2, then θ_1.

The centroid of the first half of loading displaces vertically by $(L/2)\theta_1$. The second half displaces the same from θ_1 being equal to θ_2. We can now write our work-energy equation:

$$2W(L\theta_1/2) = M_P \left[\underbrace{\theta_1}_{B} + \underbrace{\theta_1 + \theta_2}_{C} + \underbrace{\theta_2 + \theta_3}_{D} \right] \quad \rightarrow \quad \frac{WL}{M_P} = \frac{2h}{L} + 4. \qquad (18.6)$$

This is smaller than the previous result by an amount of h/L, giving us our true collapse load. Asymmetrical collapse modes in general have fewer hinges by their nature compared to their symmetrical counterparts and tend to dissipate less energy. Consequently, there is less external work and a smaller collapse load.

Both results also converge when the roof is flat, $h = 0$, despite the first case having an extra hinge. Without the pitch, there is no splaying in both cases and only the upper three hinges are active, giving equal collapse loads.

18.5   Final Remarks

The elastic bending moment profile provides a reasonable starting point for assessing positions of extreme bending, which seed plastic hinges. Their number is equal to the number of redundancies plus one for regular collapse of the entire structure but can be less if confined to part of the frame, which is highly likely in practice for more detailed structures.

Candidate positions for hinges are typically corner junctions, points of application of force, or built-in supports and feet; with distributed loadings, the span-wise hinges are not always in the middle. The most likely collapse mechanisms are beam-, sway- and combined collapse, with hinge positions linked to where salient bending moments occur.

The rotations thence displacements for collapse motion stem from the instantaneous centres of rotation for each moving body within the mechanism; plastic hinges connect undeformed body parts undergoing compatible displacements, which project from a common point of rotation. All kinematical quantities are taken to be small, in order to give invariant collapse loads that do not depend on how far their displacements have progressed.

The external work done centres on calculating the displacement of either points of application of forces or the centroidal force position of distributed loading on a particular moving part. Finally, we seek the smallest collapse load, or we plot an interaction diagram to compare between modes. When our collapse mechanism is the true mode, the loading is just on the brink of its safe limit; collapse and equilibrium co-exist only because we have also carried out a Lower Bound assessment.

19 When Buckling Occurs: And not

Buckling of a structure is signified by sudden changes in load capacity and expected deformation. It can equate to premature failure, of a geometrical type and not of material degradation (the latter, however, may promote the former). In order to predict its onset, we must consider equilibrium in the deformed, buckled configuration and, for relative simplicity, confine ourselves to shallow gradients and small displacements.

Despite revealing the critical buckling load, this analysis cannot say anything about the geometrical degree of buckling. This outcome can prove confounding, but it stems from the inherent 'perfect' treatment of the structure. In practice, real structures are not perfect and their buckling is not sudden; and paradoxically, everything about load and deformed shape is output from their analysis. The following examples try to clarify these important differences, not only in terms of their behaviour but of our approach to matters, and to set the tone for buckling design.

19.1 Perfect Buckling

Consider end-wise compression of the cantilever shown in Fig. 19.1(a). Experience tells us that it will ultimately give way laterally, *orthogonal* to the direction of compression. The axial force, P, performs work only if it can move its point of application towards the base. But, as we have adopted before, direct axial strains are negligible compared to those from bending – when curving and extension co-exist.

Work is only realised, therefore, from lateral curving: if we imagine the length along the displaced beam to be equal to L, its length projected onto the original straight cantilever is smaller than L. This gives us our favourable end-wise displacement for P, whose line of action also displaces vertically (but which does not contribute to work done).

The base reactions are a moment, C, and axial force, also P, in Fig. 19.1(b). The latter is always 'active' but C is non-zero only from deformation, which we set equal to $-P\delta$ for its specified direction, where δ is some vertical displacement of tip upwards: note that there are no vertical forces.

An orthogonal coordinate system has origin at the base, where the x-axis runs along the perfectly straight, stress-free cantilever. The coordinates of a point of interest on the deformed beam centre-line are (x, v), with v measuring its vertical displacement: any axial displacements are negligible compared to v when the overall deformation is small.

Moment equilibrium of the associated free body tells us that $M - C - Pv = 0$, where M is the positive bending moment on the right-hand side. This is equal to the bending stiffness, EI, times the usual curvature of the beam centre-line equal to $-\mathrm{d}^2v/\mathrm{d}x^2$ for the positive direction of v. Replacing C and M, and rearranging, we have our governing equation of deformation:

$$EI\frac{\mathrm{d}^2v}{\mathrm{d}x^2} + Pv = P\delta \quad \rightarrow \quad \frac{\mathrm{d}^2v}{\mathrm{d}x^2} + \alpha^2v = \alpha^2\delta \qquad (19.1)$$

after dividing both sides by the bending stiffness and setting $\alpha^2 = P/EI$.

In solving this, we recognise that δ is part of the solution, *i.e* $v = \delta$ at $x = L$; the end displacement axially is assumed to be insignificant compared to L, which stipulates this boundary condition at the original end of the cantilever.

There are also two other boundary conditions, of zero displacement and zero gradient at the base. The general solution, which combines the complementary function and the particular integral, can be written in terms of unknown constants, A and B, as:

$$v = A\cos\alpha x + B\sin\alpha x + \delta. \qquad (19.2)$$

The first boundary condition sets A equal to $-\delta$; the second, after differentiating, $B = 0$: thus, $v = \delta(1 - \cos\alpha x)$. The displacement profile is now known except for its amplitude δ, but from v we find $\delta = \delta(1 - \cos\alpha L)$: or $\cos\alpha L = 0$.

We cannot actually solve for δ, and the end-wise displacement is unspecified even though the general profile exists. However, we can find non-zero solutions for the second, *transcendental* equation, which are $\alpha L = \pi/2, 3\pi/2....$ Recalling our definition of α, then $P/EI = \pi^2/4L^2, 9\pi^2/4L^2$ *etc.*, and the critical force at buckling will be the lowest value: $\pi^2EI/4L^2$, which we denote as P_C.

Higher values represent higher-order variations of mode shape; for example, the second one sets $v = \delta(1 - \cos 3\pi x/2L)$. This, and the rest, are theoretically possible but unlikely in practice because we would have to prevent, somehow, key displacements and rotations associated with the other mode shapes.

Our final solution, therefore, tells us about the critical loading and the buckled profile but not its amplitude. This absence stems from the transcendental nature of solution: that the critical value of P is independent of δ; rather that δ does not depend on P. But in thinking about this aspect of structural behaviour, we have moved beyond buckling into the *post-buckling* phase.

Small-displacement equilibrium enables us to find the critical transition from the ordinary response, of axial compression, to buckling out of plane, and not beyond. A mode shape is essential for analysis but its amplitude does not matter; indeed, we may argue that we have our critical buckling load when δ is zero.

For larger displacements, our assumptions need to change to reflect them. For example, the axial force applied to the right side of the element in Fig. 19.1(b) should be split into local components applied tangentially and normally – in shear, to the element; curvature should be expressed as a rate of change of rotation (with respect

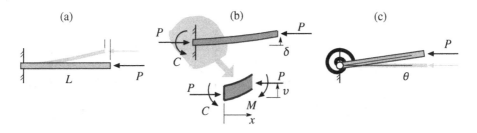

Figure 19.1 (a) Uniform cantilever compressed axially. (b) Deformed equilibrium configuration and local free body. (c) Buckling analogue via rigid bar and torsional spring.

to axial length). We now have a new, more accurate governing equation, which is not necessarily soluble in closed form nor formally demonstrated here.

These differences are expressed directly by the *discrete* analogue for cantilever buckling in Fig. 19.1(c). Bending is simulated by a linear torsional spring resisting rotations of a rigid rod about its base of stiffness, c. Under an increasing axial force, P, the rod buckles and rotates, with its line of action now displaced $L \sin \theta$ above the horizontal. Moment equilibrium about the base equates the turning effect of P to the spring moment, $c\theta$, giving $P = (c/L)\theta / \sin \theta$.

Small displacements follow from θ being small, setting $\sin \theta \approx \theta$, and P equal to c/L: we therefore arrive at a buckling threshold independent of θ, similar to the cantilever case under the same geometrical approximation.

Otherwise, the same threshold is observed for P equal to $(c/L)\theta / \sin \theta$ since the limit of $\theta / \sin \theta$ is unity as θ tends to zero. Afterwards, P increases because the ratio now increases above unity. Thus, we see a relationship between P and θ giving us a system which is stiff overall in this 'exact' phase. Conversely, the small displacement case has no stiffness because P has the same value irrespective of the value of θ: in work terms, P does no extra work (strictly within the assumption of small θ).

19.2   Imperfect Buckling

The original cantilever is also loaded now by a tip moment, m, Fig. 19.2(a.i). On first sight, we may expect similar behaviour to the case without m, where buckling at non-zero loads occurs for a perfectly straight, stress-free structure initially.

But imagine that we apply m first, then P. There will be some transverse displacements before we apply P, Fig. 19.2(b), which can exert an immediate moment at every point on the deformed centre-line. Importantly, there is *already* deformation for zero P, and we should expect a change in the complexion or character of buckling when P increases.

The analysis is as before, with the overall equilibrium setting the base moment, C, equal to $-(m + P\delta)$, with δ being the usual end-wise transverse displacement. A local free body enclosing the base allows us to write the same internal bending moment, M,

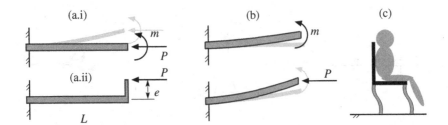

Figure 19.2 (a.i) Uniform cantilever compressed axially and carrying a tip moment; (a.ii) realisation of (a.i) loading by applying axial force at a small offset. (b) Initial loading by moment generates transverse displacements, before applying axial force. (c) Compressed and buckled legs of the chair preserve their end rotations compared to being straight originally.

equal to $C + Pv$, and the governing equation as

$$EI\frac{d^2v}{dx^2} + Pv = m + P\delta \quad \rightarrow \quad \frac{d^2v}{dx^2} + \alpha^2 v = \frac{m}{EI} + \alpha^2\delta. \tag{19.3}$$

Compared to Eq. (19.1), there is one extra term due to m and, thus, a small addition to the usual general solution:

$$v = A\cos\alpha x + B\sin\alpha x + \frac{m}{P} + \delta. \tag{19.4}$$

The same boundary conditions persist at the base, setting B equal to zero and $A = -(m/P + \delta)$. Also, $v = \delta$ at the tip returns $\delta = (m/P + \delta) \cdot (1 - \cos\alpha L)$ and for δ explicitly, we have:

$$\delta = \frac{m}{P}\left[\frac{1 - \cos\alpha L}{\cos\alpha L}\right]. \tag{19.5}$$

Now it seems that δ is defined, but recall how α and P are related, which requires us to think about its expression ultimately in either term. To illustrate as such, we consider three cases.

The first is when m and P are related proportionally, as might be contrived by applying P at a small offset, e, to the beam axis, Fig. 19.2(a.ii): the corresponding moment now applied to the tip of the beam is clearly Pe. Consequently, δ in Eq. (19.5) is governed by e, a constant, and the term $(1 - \cos\alpha L)/\cos\alpha L$.

We can plot, for example, the ratio δ/e, which is equal to the trigonometric term that follows; and we note that when $\alpha L \ (= \sqrt{PL^2/EI})$ is small, we are at the onset of loading and $\cos\alpha L$ is approximately $1 - (\alpha L)^2/2$, giving $\delta \approx \alpha^2 L^2/2 = PL^2/2EI$. The initial relationship between P and δ is thus linear.

Before plotting the full expression, however, we re-write αL differently by recalling the earlier perfect buckling load without an offset, P_C. From its expression, we can write $EI = (4L^2/\pi^2)P_C$ and, since $\alpha^2 = P/EI$, then $(\alpha L)^2 = (\pi^2/4) \cdot (P/P_C)$. Defining the force ratio P/P_C to be $\bar{p}$, then $\alpha L = (\pi/2)\sqrt{\bar{p}}$, and

$$\frac{\delta}{e} = \frac{1 - \cos(\pi\sqrt{\bar{p}}/2)}{\cos(\pi\sqrt{\bar{p}}/2)} = \bar{d}_1. \tag{19.6}$$

Our dimensionless displacement is now related directly to the applied force, and nonlinearly so.

Figure 19.3 indicates the variation of $\bar{p}$ with $\bar{d}_1$ as the black curve, which is initially linear before *softening* with an ever-decreasing gradient. The force becomes asymptotic to the value set by infinite displacement, *i.e.* when the denominator in Eq. (19.6) is zero, setting $\cos(\pi\sqrt{\bar{p}}/2) = 0$ and thus $\bar{p} = 1 (= P/P_C)$.

In the limit, the axial force approaches the buckling force of the perfect case. The moment component due to its eccentric nature always promotes a transverse deflection, enabling P to increase from zero. In this sense, there is no sudden jump in configuration and strictly no buckling here under P.

For very large displacements, the moment term, $m = Pe$, compared to Pv in Eq. (19.3) is diminished by $v \gg e$, and the loading behaves more like the perfect case but without ever reaching this limit.

The same behaviour is well known in cases where structures are naturally, if only slightly, bowed initially. Such *imperfection* means that a perfectly applied axial force already induces bending moments before deflections commence, similar to our example. The applied loading can also only approach the equivalent perfect case when displacements become large, but this is unlikely either because of plastic strains or simply that we are already violating the assumption of shallow displacements.

The second example treats m and P as independent. As before, we can replace P with $\alpha^2 EI$ in the pre-factor, m/P, in Eq. (19.5), which we expand to read as $(mL^2/2EI) \cdot (2/\alpha^2 L^2)$. Defining a new dimensionless displacement, $\bar{d}_2$ equal to $\delta/(mL^2/2EI)$, and remembering $\bar{p}$, we can write

$$\bar{d}_2 = \frac{8}{\pi^2} \frac{1}{\bar{p}} \frac{1 - \cos(\pi\sqrt{\bar{p}}/2)}{\cos(\pi\sqrt{\bar{p}}/2)}. \tag{19.7}$$

This variation is also plotted in Fig. 19.3, as the grey curve.

Note first that zero $\bar{p}$ starts off at $\bar{d}_2 = 1$, which is precisely the transverse tip deflection under a pure end moment from our list of standard results in Chapter 10: $\delta = mL^2/2EI$. The axial force rises and softens with displacement and almost merges with the first curve but not quite, as the physical conditions are slightly different: there is the same asymptotic tendency because of the dominant bending effects from Pv compared to m at larger displacements.

In the final example, m and P prevail, but now we enforce the boundary condition of zero slope at the tip. The particular solution for v and the expression for δ from earlier are valid because they satisfy the existing boundary conditions. If we now differentiate v with respect to x:

$$\frac{dv}{dx} = \left[\frac{m}{P} + \delta\right]\alpha\sin\alpha x = \left[\frac{m}{P} + \frac{m}{P}\frac{1 - \cos\alpha L}{\cos\alpha L}\right]\alpha\sin\alpha L = \frac{m}{P}\frac{\alpha\sin\alpha L}{\cos\alpha L} \tag{19.8}$$

at $x = L$. Zero gradient here obviates $\alpha = 0$, which is the unloaded case, or $\sin\alpha L = 0$. This is another transcendental equation, satisfied by $\alpha L = 0, \pi, 2\pi...$ *etc.* The lowest non-zero value is clearly π, which sets P equal to $\pi^2 EI/L^2$.

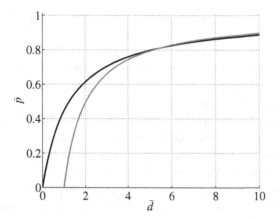

Figure 19.3 Dimensionless axial force *vs* transverse tip displacement for a compressed cantilever carrying an end moment: black ($\bar{d}_1$, Eq. (19.6)), when the applied force and moment are coupled; grey ($\bar{d}_2$, Eq. (19.7)), when they are not.

There are three points to note. The critical buckling load is larger than P_C because the tip is further constrained, which increases the overall axial capacity. It is exactly four times larger because the buckled mode shape is anti-symmetrical about the middle of the cantilever: up to this point, we can excise a free body cantilever loaded axially by P but no end moment because the bending moment is zero at $x = L/2$. The buckling load, therefore, has the same expression as P_C where L is now replaced by $L/2$.

Second, there are no solutions for which $P < 4P_C$ and buckling is instantaneous as per the perfect case without m. Consequently, the end displacement δ reverts, again, to being undefined in absolute terms. Of course, it is defined in terms of m and P from Eq. (19.5), noting that $\cos \alpha L = -1$ when $\alpha L = \pi$ and returning $\delta = -2m/P$. Plugging into v, however, we merely confirm $\delta = \delta$.

Third, that δ *is* negative, and the displaced mode shape therefore lies below the original cantilever axis. The axial force in this instance will tend to curve the beam downwards whilst positive m bends the beam upwards, nullifying any negative tip gradient induced by P. Indeed, it is this *antagonism* between the effects of P and m upon the bending moment profile which gives rise to sudden buckling.

On a frivolous note, this analysis can describe the compressive capacity of the slender chair legs shown schematically in Fig. 19.1(c). Compression is due to the seated weight of the user (and of the chair), and the horizontal plate tying the vertical front and back legs together enforces zero end rotation at both of their tops (or tips). We have assumed their feet are similarly tied to foist the same zero gradient at the bottom (bases). The maximum weight, without a safety factor, should be less than $4P_C$ times the number of legs.

19.3   Final Remarks

Small loading adjustments can profoundly affect the response of a structure susceptible to buckling. The level of change in the governing equation of deformation is equally innocuous but its solutions can be radically different. There is either progressive loading and emergent quantifiable displacements or a sudden jump to a buckled shape of indeterminate amplitude. Boundary conditions play a key role, and we shall return to their effects explicitly in the next chapter.

Notwithstanding, perfect buckling overestimates the capacity of the structure compared to the imperfect case: and all structures, no matter how well made, will have small geometrical imperfections – usually within acceptable tolerances. All structures must deform progressively during 'buckling' with no single, well-defined buckling load as per the perfect case, which is too high.

Designers, however, operate better with clear limits. One expediting route, as we saw for elastic stiffness in Chapter 14, is to extract the limit of *elastic* loading as it builds up, when the cross-section first begins to yield. This returns a single, pseudo-buckling load conveying the (elastic) capacity of the slightly imperfect structure in compression, which may be confirmed with any of the previous examples.

20 The Nature of Loads

In Chapter 2, we considered applied loads whose size depends on the displacements they induce. Although not explicitly stated, both quantities are *well behaved*: displacements change incrementally for relatively small changes in the loading, and *vice versa*.

Imagine, however, when small changes in the equilibrium configuration can drastically alter their profiles. An everyday example concerns sitting on a four-legged chair on an uneven, hard floor. Rocking from side-to-side by just a few millimetres can result in some force or no force at all in the non-pivoting legs.

Similarly, a square block being 'rolled' from one edge to the next will transfer its weight between these edges just as one face makes and loses contact, almost immediately. Fortuitously, both situations are statically determinate provided we know the precise configuration of the system.

Statical indeterminacy, however, compounds the uncertainty. For example, the two-span beam in Fig. 20.1(a) carries a uniform loading across three equi-spaced simple supports. We expect the response to be symmetrical for a perfect layout and no initial stresses (*i.e.* self-stress). Furthermore, underneath a *rigid* beam, only the central reaction is non-zero and equal to its weight; or, the outer reaction forces are equal to half of the weight, but nothing else. Both cases are verified from moment equilibrium.

When designing each support for maximum strength, we would assume a full weight capacity for the middle together with half for the outer legs despite separate responses in theory. If the right-most support, say, is in fact detached, Fig. 20.1(b), only the middle reaction is ever fully loaded: statical determinacy precludes any left-side reaction force. It is possible for the beam to 'rock over' over onto the right-side support if the left one permits lift-off, but this does not change the middle reaction.

Elasticity, however, prevents the beam from being in a perpetual state of reaction force indeterminacy. It may deflect far enough to come into contact with the support, thereby generating a reaction and statical indeterminacy, which alters the size of the others. Paradoxically, a larger gap prevents a reaction but may permit large enough displacements and thence their gradients that disrupt the symmetry of the self-weight loading.

There is some evidence (anecdotal) to suggest similar, if disastrous, consequences for the transportation of large stone blocks in Classical Antiquity. Each block is supported on two logs acting as wheels, and, for a large enough span, may crack underneath in the middle, Fig. 20.1(c). To counteract this tendency, workers were tempted to add a third, middle log for a better distribution of the block weight; but if they do,

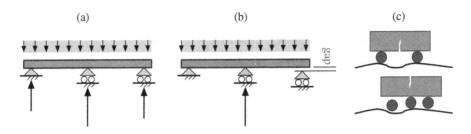

Figure 20.1 (a) Uniformly loaded, indeterminate two-span beam with three unknown support reaction forces. (b) Detached right-side support may lead to a single reaction force. (c) Large stone supported on a pair of wheels over rough ground, fractures in the middle of the underside: a third correcting wheel added to the middle results in cracking at the top.

the cracking simply flips position to the top of the block, Fig. 20.1(c), with no possible improvement in failure rates.

The ground beneath is never perfectly flat and the inflexible block only ever makes contact with two logs, either the outer pair or one of them and the middle log. Assuming a uniformly distributed mass, the maximum bending moment on the central cross-section in both cases is the same but of opposite signs. Stone fractures under tension because it is a ceramic and ultimately the top or bottom layer is most susceptible, and equally so for both contact conditions.

20.1   Rucking Behaviour

A *ruck* forms in a rug when part of it is pushed against the rest, accidentally or otherwise. It remains, annoyingly, in place for a rough-enough floor beneath despite gravity and a desire to be straight (from unbending). A simple model for ruck equilibrium is shown in Fig. 20.2(a), which captures the rug as a slender elastic beam in cross-section of length L. Initially, the beam is flat, uniform and free of stresses.

The middle region of the beam is detached from the rough floor beneath, having been displaced upwards and symmetrically; this 'hump' is the basic shape of the ruck. It is connected smoothly to flat, equal portions on both sides, which have pulled in slightly, and the overall self-weight is a uniform loading intensity, w. Equilibrium is maintained by frictional forces distributed over the flat parts, which resist their outward tendency to slip as the ruck flattens.

We are interested in various features, most notably, the nature of the forces applied to the beam and how they *effect* displacements; and the relationship between the contact width, λL, and the coefficient of friction of the floor, μ, for limiting behaviour. The final shape is obviously governed by deformed equilibrium conditions, where we can be tempted to ignore the progress of deformation. The latter, in fact, turns out to be rather crucial in our approach.

Figure 20.2(b) shows a free body of the left-side flat part. The beam centre-line does not displace or curve, so there is no bending moment or shear force accordingly.

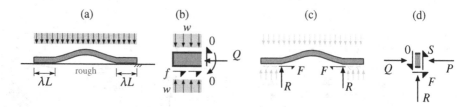

Figure 20.2 (a) Symmetrically deformed uniform elastic beam, simulating a raised ruck and smoothly connected flat sides resting on a rough floor. (b) Equilibrium of a beam element in the left side. (c) Reaction forces where the ruck joins the sides. (d) Corresponding equilibrium at these junctions (left-side).

The normal contact reaction can only be equal to the self-weight, giving a limiting and uniform distributed friction intensity, f, equal to μw. The axial force in this part of beam, Q, balances the total friction force up to that point, increasing linearly from zero up to $\mu w \lambda L$ where contact is first lost.

The total normal reaction is $2w\lambda L$, which clearly does not balance the self-weight, wL. No other vertical forces can be applied over the flat region because they would counter the equilibrium we have just discussed, and none can prevail where the ruck has lifted off because there is no contact.

At the *junction* of these two configurations (on both sides, of course) we reconcile matters with a pair of vertical point forces, R, applied to the beam, Fig. 20.2(c), each equal to one half of the weight of the ruck, $wL(1-2\lambda)/2$. These extra normal reaction forces come with their own limiting frictional forces, F, equal to μR.

At the lift-off location, there are now discontinuities in both the shear force and the axial force, Fig. 20.2(d). The latter increases to P equal to $Q + F$, which is $\mu w \lambda L + \mu w L(1 - 2\lambda)/2 = \mu w L/2$, *i.e.* one half of the beam weight multiplied by μ. The bending moment, on the other hand, behaves smoothly because the deformed geometry is continuous.

We can now isolate the ruck as an externally loaded free body in Fig 20.3(a). This bears the hallmarks of a *buckled* beam under end-wise axial forces P and transverse uniform intensity w, which gives rise to the vertical reactions, R. As well as zero displacements at both ends, there is zero local gradient for continuity with the adjacent flat regions.

The beam has buckled to favour out-of-plane movement during its original compression. Furthermore, the buckling analogy is structurally equivalent in view of the shape and loads persisting at the point of slipping; if the buckled equilibrium is not viable, neither is limiting equilibrium of the deformed ruck. We can, therefore, think of applying the force P to the original ruck *if it were flat* and increasing from zero until we obtain the buckled shape we see.

The transverse loading, however, presents a subtlety. If w acted (somehow) in the opposite sense, it would complement the actions of P by promoting transverse displacements from the outset: P, itself, would simply add to their magnitude with immediate and progressive deflections, without a sudden buckling jump, *c.f.* Chapter 19.

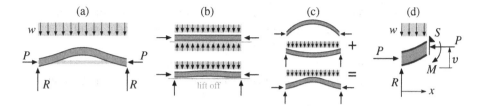

Figure 20.3 (a) External forces applied to the ruck in Fig. 20.2. (b) Distributed floor reaction changes to point forces when buckling starts. (c) Uniform self-weight counters buckling displacements with zero displacement and rotations at both ends of beam. (d) Free body of a part of ruck from its left end.

Presently, however, P and w are antagonistic because of the 'hard' floor preventing downward displacements. Moreover, the effect of w on the displaced shape is not rightly felt until there is lift off, Fig. 20.3(b), when the distributed reaction from the floor disappears – when buckling has *already* occurred.

It is tempting to say that the critical value of P can, therefore, be found by ignoring w but the latter contributes to the requirement for zero gradient at both ends. Imagine applying P without w: we have buckling of a classical Euler strut whose pin-ends are free to rotate to some inclination, Fig. 20.3(c). The effect of w is to 'unbend' the already buckled beam, restoring its end gradients to zero in the process. The interplay between P and w is thus clear, and no further insight can be gleaned without a formal analysis.

Part of the deformed ruck is shown as a free body in Fig. 20.3(d), which starts just to the right of R, where the left-side shear force is also R, the axial compressive force is P and bending moment is zero. The origin of a Cartesian coordinate system is located at the left on the beam centre-line, with x horizontal and v capturing small vertical displacements above the floor, as shown.

On the right side cross-section there is a positive shear force S and bending moment M, as well as P, which remains horizontal from the gradients being shallow. Resolving vertically, we find $S + R - wx = 0$, which expands to $S = w(x + \lambda L - L/2)$, and taking moments about the origin; $M - Pv + (wx) \cdot (x/2) - Sx = 0$.

The bending moment is thus generated twofold as we expect, from the leverage vertically of P and horizontally of w and S (and R). We may replace M with the generalised Hooke's Law for bending, remembering that the beam centre-line curvature is equal to $-d^2v/dx^2$ from our positive directions of v and M. The governing equation of deformation is now:

$$EI\frac{d^2v}{dx^2} + Pv = -wx\left[\frac{x}{2} + L\left(\lambda - \frac{1}{2}\right)\right]. \tag{20.1}$$

The right-hand side defines a quadratic particular integral, and we first divide both sides by the bending stiffness EI before writing $\alpha^2 = P/EI$. Consequently

$$v = A\cos\alpha x + B\sin\alpha x - \frac{w}{P}\left[\frac{x^2}{2} + xL\left(\lambda - \frac{1}{2}\right) - \frac{1}{\alpha^2}\right] \tag{20.2}$$

can be shown to be a valid general solution when substituted into Eq. (20.1), where the constants A and B are found from boundary conditions of $v = 0$ at $x = 0$ and $x = L(1 - 2\lambda)$, the right-side end of the ruck.

The first condition yields $A = -w/P\alpha^2$: if we define $\theta = \alpha L(1 - 2\lambda)$, the second condition sets $B = -(w/P\alpha^2) \cdot (1 - \cos \theta)/ \sin \theta$. Our formal solution for v is thus:

$$v = \frac{w}{P}\left[\frac{-\cos \alpha x}{\alpha^2} + \frac{\sin \alpha x}{\alpha^2}\frac{(\cos \theta - 1)}{\sin \theta} - \frac{x^2}{2} - xL\left(\lambda - \frac{1}{2}\right) + \frac{1}{\alpha^2}\right]. \qquad (20.3)$$

This (formidable) expression defines how the displacement varies according to P and w (as well as EI, L and λ, which are maintained in α and θ). In essence, it reflects quantitatively most of our earlier thought experiment in Fig. 20.3(c) without considering zero gradients at both ends of the ruck: we have also yet to find the 'critical' value of P, which just maintains this complete buckled shape.

We therefore differentiate the above with respect to x to produce

$$\frac{dv}{dx} = \frac{w}{P}\left[\frac{\sin \alpha x}{\alpha} + \frac{\cos \alpha x}{\alpha}\frac{(\cos \theta - 1)}{\sin \theta} - x - L\left(\lambda - \frac{1}{2}\right)\right], \qquad (20.4)$$

which is then set equal to zero at either $x = 0$ or $x = L(1 - 2\lambda)$: the same expression is returned either way. Before setting equal to zero, however, we substitute for α from $\theta/L(1 - 2\lambda)$ (or the other way round) for consistency in either variable. We then produce a right-hand side in terms of θ alone by re-arranging for a dimensionless term on the left, which we denote as $\bar{g}$

$$\bar{g} = \frac{dv}{dx} \cdot \frac{P}{wL(1 - 2\lambda)} = \frac{\cos \theta - 1}{\theta \sin \theta} + \frac{1}{2}. \qquad (20.5)$$

Note that the term $P/wL(1 - 2\lambda)$ is the ratio of axial force to ruck weight.

Values of θ which admit zero $\bar{g}$ are tantamount to fulfilling zero gradient, but they also express the critical buckling force because of how θ is defined in terms of α and other variables. However, the right-hand side cannot be solved in closed form despite its relative simplicity.

Instead, we plot $\bar{g}$ in Fig. 20.4(a) in terms of θ, which must be positive, and inspect the variation closely for zero values. The first occurrence is when $\theta = 2.86\pi$ and next, 4.92π, and so forth beyond the indicated variation. Each new value of θ increases the *frequency* of trigonometric terms in Eq. (20.3), which changes the character, or *mode shape*, of the deformed profile, as we shall see. There are also asymptotes at θ equal to π, 3π etc. where the gradient is not defined.

Our solution is now complete, and the lowest buckling force is thus $P = (2.86)^2\pi^2 EI/L^2(1 - 2\lambda)^2$. The beam properties define EI, and λ controls the extent of the ruck shape. From the earlier balance of horizontal force, recall $P = \mu wL/2$, which defines the minimum friction coefficient. For example, assuming a unit depth of beam and thickness t, w is equal to its density, ρ, multiplied by tg; I is $t^3/12$, and:

$$\mu \geq \frac{(2.86)^2\pi^2}{6} \cdot \frac{E}{\rho g L} \cdot \frac{t^2}{L^2} \cdot \frac{1}{(1 - 2\lambda)^2}. \qquad (20.6)$$

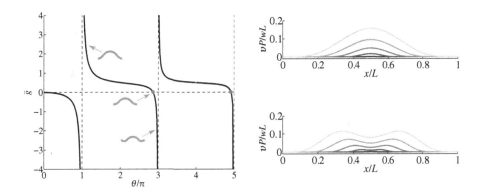

Figure 20.4 (a) Variation of dimensionless gradient via Eq. (20.5) with axial force. Each schematic inset shape of the ruck has different end rotations; grey dots are load cases with zero end rotations and equivalent to the buckled cases. (b.i) Dimensionless shapes of ruck and sides via Eq. (20.3) when $\theta = 2.86\pi$ at the lowest buckling load; (b.ii.) the same when $\theta = 4.92\pi$ at the next highest buckling load. In both, the ruck extent is 90% of L (light grey), stepping down 20% each time to a final 10% (black).

Alternatively, for a given proportion (t/L) of beam, we can adjust the ruck extent through λ until the sides begin to slip, when μ is matched by the actual coefficient of friction: this gives a simple practical way for finding frictional properties.

Note that μ becomes very large as λ tends to its limit of one-half in the above formula, suggesting relatively bigger rather than smaller rucks. If smaller, then the buckling force becomes disproportionately large from its reciprocal dependence on the square of the ruck length.

For confirmation of the ruck shape, we plot Eq. (20.3) in Fig. 20.4(b.i) after making v dimensionless by multiplying both sides by P/wL. The limits of axial position, x/L, are now zero and unity, and we specify different ruck sizes by selecting λ in the range 0.05 to 0.45 in steps of 0.1, giving a ruck span of 90%, 70% ... 10% of L, respectively. Unexpectedly, a taller ruck is a wider ruck, but we now see similar height-to-width ratios each time.

At the next highest buckling force solution ($\theta = 4.92\pi$), the ruck shape begins to differ – our higher mode shape. Using the same conditions, Fig. 20.4(b.ii) shows a middle 'dimpled' region in which the ruck curves downwards, without touching the floor. This also fits with our experiences, provided the floor is rough enough: we may form an ordinary ruck before pressing down its middle into a second resting shape as in (b.ii).

20.2 Snapping Through

A more emphatic, rapidly changing load *response* comes from a pin-jointed arch *snapping-through*. In Fig. 20.5, a centrally applied vertical force, P, pushes the arch

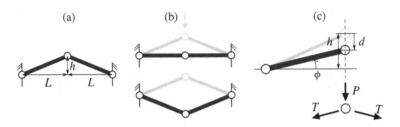

Figure 20.5 (a) Initial geometry of a symmetrical and shallow pin-jointed arch. (b) Alternative load-free configurations. (c) Displacement of central pin-joint and nodal equilibrium.

into compression well beyond the bounds of small displacements – without material yielding, of course. In theory, P is not a passive load but can change its value to mete the displaced shape exactly. This is usually not possible in practice for simple loadings, and we discuss later what happens instead.

The arch is the familiar three-pinned truss with a base width of $2L$ and initial shallow 'rise', h, Fig. 20.5(a). It is statically determinate – at least for small displacements, *c.f.* Chapter 4, but for larger, we note two extra, soluble equilibrium configurations with no load applied, see Fig. 20.5(b).

The first occurs when the bars are horizontal with no vertical component of bar force acting on the central pin-joint. Each bar has decreased in length from $\sqrt{L^2 + h^2}$ to L giving a compressive strain approximately equal to $(1/2) \cdot (h/L)^2$ when $h \ll L$ using the Binomial Theorem. The current compressive bar force multiples this strain by EA for a cross-sectional area, A.

The second state is a vertical reflection of the first: the inverted bars have the same initial length, they are not strained and, without bar forces, P is zero. In moving from the initial state through horizontal to being inverted, strain (and stress) is accumulated before being relieved in both bars.

Because the central pin displaces up to $2h$, we must assess equilibrium of the *displaced* truss and its geometrically nonlinear response. Our analysis mimics that for our earlier buckling problems in Chapter 19, but we do not expect physically similar responses because the truss deforms *progressively* as displacements become relatively larger; there is no switch to a different mode shape. There is, however, a type of buckling, depending on how the loading is controlled, as we shall see.

The central pin-joint moves d vertically downwards, Fig. 20.5(c). Each bar is currently inclined at a shallow angle ϕ to the horizontal where $\tan \phi = (h - d)/L \approx \phi$. Nodal equilibrium returns $P = -2T \sin \phi$ for positive tensions, which can be rewritten $P \approx -2T\phi = -2T(h - d)/L$.

The corresponding tensile strain, ϵ, is determined from the change in bar length divided by its original length:

$$\epsilon = \frac{\sqrt{L^2 + (h - d)^2} - \sqrt{L^2 + h^2}}{\sqrt{L^2 + h^2}} \approx \frac{1}{2} \frac{(h - d)^2}{L^2} - \frac{1}{2} \frac{h^2}{L^2}. \qquad (20.7)$$

Once again we deploy the Binomial Theorem and retain terms up to, but not beyond, second-order (quadratic) variation for simplicity.

Multiplying the strain by EA, we replace T in our force equilibrium statement to reveal P; in doing so, recognise $(h - d)^2 - h^2$ as the difference of two squares now written as $-d(2h - d)$. Our final expression is therefore:

$$P = \frac{EA}{L^3} \cdot d \cdot (h - d) \cdot (2h - d). \tag{20.8}$$

Its roots have been deliberately highlighted; $d = 0$, $d = h$ and $d = 2h$, i.e. the initial, horizontal and inverted load-free states. At the onset of loading, d is relatively small and $(h - d) \approx h$, and so forth, to give an initial force that depends linearly on displacement, i.e. $(2EAh^2/L^3)d$.

All cases collapse onto a single curve plotted in Fig. 20.6 when we define $\bar{P} = PL^3/EAh^3$ and $\bar{d} = d/h$, setting $\bar{P} = \bar{d}(1 - \bar{d}) \cdot (2 - \bar{d})$. The 'up-down-up' profile clearly evinces a cubic response, which crosses the horizontal axis at our load-free values, highlighted by the grey circles.

Following this equilibrium path as the displacement increases presumes that the applied force *complies* precisely. From zero, we follow the curve up to point A straightforwardly, despite a decreasing gradient and a diminishing stiffness. Just beyond A, the force must begin to decrease: we now have to remove weights from a scale-pan or withdraw pressure from a hydraulic jack – if this is how the loading is practically applied.

But this has to be perfectly coordinated with the conditions at point A, where $\bar{d} = 1 - 1/\sqrt{3}$ from solving $dP/dd = 0$ for stationary points. If the load is reversed too early, we move back towards the origin; if not, we cannot follow the curve in a smooth fashion past A. Equilibrium, however, has to be upheld, resulting in a 'jump' across the plot to the same value of force on the other side at point B. Because our displacement is now $\bar{d} \approx 2.15$, the truss has more than inverted ($\bar{d} = 2$).

Our equilibrium jump manifests as a dramatic change in configuration for an imperceptible change in loading; our introductory rocking chair behaved quite oppositely, c.f. slight motions giving rise to an immediate transfer of reaction forces. Here the jump is often rapid if the truss is lightweight, and marked by an audible 'snap'. Such behaviour is akin to buckling at constant load only because the loading cannot 'anticipate' how displacements progress: in that sense, our experiment is said to be *force controlled*.

The applied force at B can be increased to move further up the rising curve, or decreased until it changes direction, causing the truss to revert progressively until equilibrium reaches point C at $\bar{d} = 1 + 1/\sqrt{3}$. If $\bar{P}$ is reduced, even incrementally, the truss snaps back to point D at $\bar{d} \approx -0.15$, with the negative force ($\bar{P} \approx -0.385$) now pulling the truss upwards above its initial shape.

If we can marshal, instead, increments in displacement at the point of application of the loading for whatever value of loading is required, we progress along the entire curve continuously. Such *displacement controlled* testing is enabled nowadays by certain electro-mechanical tensometers.

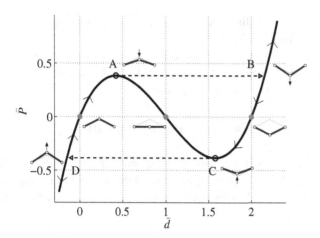

Figure 20.6 Variation of dimensionless vertical force *vs* displacement of the central pin-joint of a shallow arch. Inset figures show deformed shape of arch: grey solid circles are load-free equilibria; and paths A aato B and C to D mark jumps in configuration during force-controlled loading.

Moving, say, from point A to C along the curve in Fig. 20.6, the gradient, and hence stiffness, is always negative. Our intermediate load-free state at $\bar{d} = 1$ is not a stable one because any slight perturbation away from it is not resisted by an increasing force in the same direction. By the opposite argument, those states at $\bar{d} = 0$ and $\bar{d} = 2$ are definitely stable.

The truss is, of course, an idealised description of the snap-through phenomenon. But it is trivial to contrive in practice: slicing off just less than the top half of a tennis ball produces a continuous form of the arch or 'cap', which can be readily inverted by manually pushing through. There is now bending as well as compression of the cap material, which simulates the deformation of the truss bars, and the cap edge, which does not easily expand or contract, captures the boundary constraint of the truss supports. Because we have two distinct and 'content' load-free states, these structures are said to be *bistable*.

20.3   Final Remarks

As in buckling problems, the deformed equilibrium is important; similar to the problems in Chapter 2, the size of the applied loads depend on the displacements they cause. All are statically indeterminate, and there will be an element of geometrical nonlinearity. Such specialised problems are rare in traditional structures applications: the loading is often *invariant* because displacements tend to remain very small. But we see a new relationship emerge between the shape of the structure and the loading, which may be exploited favourably in view of other applications outside the Civil Engineering realm, as we shall now see.

21 Hooke and Heat

When heated, most materials expand in size, with cooling producing a contraction. The level of expansion (or contraction) for a rise (or drop) in ambient temperature is not usually challenging for how it is used ultimately unless the temperature change is extreme, which may also compromise the integrity of the material. If the expansion (or contraction) is prevented instead, high internal forces can develop, leading to other types of failure, for example, railway lines buckling sideways from straight during a severe heat-wave.

Such unwelcomed distortion also tells us something about the amount of temperature change. This relationship is exploited in *reverse* in the design of a *bimetallic strip*, which becomes curved when deliberately heated. Being small, it is part of a thermostatic switch found inside virtually every household kettle. When subjected to the boiling water conditions, the strip distorts sufficiently to break the electrical heating circuit, preventing the kettle from 'boiling dry'.

Understanding the relationship between curving and heating of the strip requires us, first, to modify our constitutive material law of stress and strain to include a thermal response. For a linear elastic material, we simply add a *thermal strain* component, Λ, to the usual one-dimensional equation employed during beam analysis:

$$\epsilon = \sigma/E + \Lambda. \tag{21.1}$$

The usual elastic stress is σ, with E as the Young's Modulus; Λ is calculated by multiplying the (linear) coefficient of thermal expansion, denoted usually by α, by ΔT, the rise in temperature in Kelvin. The resulting change in geometry due to mechanical *and* thermal effects is the observable strain, ϵ.

Much is conveyed by this simple expression. If there is no stress, the material expands 'freely', without constraint, and we see Λ $(= \epsilon)$ directly. If the strains are completely inhibited ($\epsilon = 0$), say, by confining the material inside an exactly-fitting rigid box, $\sigma = -E\Lambda$, causing compressive stresses during heating. Otherwise, straining is some combination of both components.

21.1 Bimetallic Strip Behaviour

From its namesake, a bimetallic strip is made by joining two thin layers of different metals, which attempt to expand differently whilst curving in unison. Moments applied to an ordinary strip or beam usually induce curving, but here they are zero without loading, which does not prevail presently: the strip freely curves.

Such paradoxical behaviour is further expounded by the nature of the elastic strain distribution through the total thickness. We know from ordinary beam curving that geometrical compatibility demands a linear variation about a neutral plane, or axis, in side view. Ambient heating produces uniform thermal strains, so linear strains are coupled to linear stresses from Eq. (21.1) and *vice versa*. And linear stresses tend to produce bending moments, even though we expect none.

To resolve matters, we note that the cross-section is individually rectangular for each layer. Bending of similar 'composite' beams is usually tackled first by *transforming* the bimetallic cross-section into a single material by widening (or reducing) the cross-sectional width of one of the layers in accordance to ratio of their Young's Moduli. The traditional neutral axis is now located at the centroid of the transformed cross-section, which is both unstrained and unstressed; thermal strains presently, however, undermine the reciprocity of this assertion.

The key to solving the problem is *not* to transform the section and to assume a general linear strain distribution without assigning a particular neutral axis. Equilibrium, compatibility and generalised Hooke's Law in both layers now have to be considered from scratch. But we might ask first why should the strip bend at all when heated?

Each constituent layer is separated first in Fig. 21.1(a). Their lengths are L and they have the same thickness, $t/2$, for convenience. The top layer has a Young's Modulus E_T and a thermal strain component Λ_T ($= \alpha_T \Delta T$); the bottom, E_B and Λ_B.

Uniform heating produces different free extensions of each layer, Fig. 21.1(a), denoted as e, with $e_T = \Lambda_T L$ and $e_B = \Lambda_B L$. When joined together and heated again, there can be no step change in strain at the common junction layer, foisting a different yet compatible strain response through the thickness.

Without bending, the strains through each layer must be the same, setting a very specific relationship, $\sigma_T/E_T + \Lambda_T = \sigma_B/E_B + \Lambda_B$, using Eq. (21.1). Each layer experiences an axial force, $\sigma wt/2$, but if no net force is applied overall, these forces and hence stresses must be equal and opposite.

Without solving for σ_T or σ_B, however, these forces always produce a net bending moment, and if none is applied externally during heating, we violate equilibrium.

The strip must therefore curve, Fig. 21.1(b). Because each layer tends to extend as well, the neutral axis is still not obvious, but we choose the junction layer conveniently to be our reference axis for 'centre-line' curving.

We cannot discount bending moments altogether because the strip may in fact be loaded transversely during its operation. Recall that in ordinary beams, moments and curving are connected by the generalised Hooke's Law for bending of a uniformly curved element of beam.

Here, we can think of applying equal and opposite bending moments, M, *after* heating, Fig. 21.1(c), which increases (or decreases) an already curved centre-line. The generalised Hooke's Law for the strip must, therefore, couple the curvature κ of the centre-line, M, and the level of heating, as we shall see.

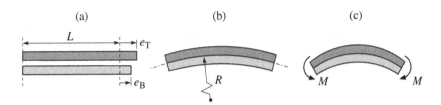

Figure 21.1 (a) Different thermal expansions of two strips made from different metals under a rise in ambient temperature. (b) The same strips, now connected to form the two layers of a single bimetallic strip, also curve in response to heating. (c) Moments applied to the deformed arrangement.

The linear strain distribution in Fig. 21.2(a) has a net centre-line value equal to ϵ. At a general height z above it, a given fibre has axial strain $\epsilon + z\kappa$, with positive values of κ giving downwards curving.

The general form can be re-cast as a uniform strain and a pure bending component for each layer as shown in Fig. 21.2(b). The former are equal to the original levels halfway through each layer, *i.e.* $\epsilon \pm (t/4)\kappa$.

Let there be corresponding axial forces, P_T and P_B, acting on each layer cross-section, Fig. 21.2(c). These multiply the average stress by the strip area, which can be determined from the uniform strain levels and Eq. (21.1) as follows:

$$P_T = E_T \cdot w(t/2) \cdot \left[\epsilon + (t/4)\kappa - \Lambda_T\right], \quad P_B = E_B \cdot w(t/2) \cdot \left[\epsilon - (t/4)\kappa - \Lambda_B\right]. \tag{21.2}$$

Each layer also carries separate bending moments, M_T and M_B, which are related purely to curvature by the 'EI' of each layer:

$$M_T = E_T \cdot \frac{w(t/2)^3}{12} \cdot \kappa, \quad M_B = E_B \cdot \frac{w(t/2)^3}{12} \cdot \kappa. \tag{21.3}$$

There are no thermal strain terms because each local moment is derived from linear strains about their own neutral axis, as shown in Fig. 21.2(b).

No axial force is applied to the strip, but there can be a net bending moment, M, over the entire cross-section to satisfy equilibrium ultimately with some applied transverse loading. From force and moment equilibrium, we can write:

$$P_T + P_B = 0; \quad M_T + M_B + (t/4)[P_T - P_B] = M. \tag{21.4}$$

The first condition specifies the relationship between ϵ and κ in terms of the thermal properties after substituting Eq. (21.2): the amount of axial lengthening *vs* curving are thus coupled. Setting the modular ratio to be $E_T/E_B = m$, we find:

$$\epsilon = \left[\frac{1-m}{1+m}\right] \cdot (t/4)\kappa + \frac{m\Lambda_T + \Lambda_B}{1+m}, \tag{21.5}$$

which can be written explicitly in terms of κ if we like.

This is a non-trivial expression because of m despite our simplified arrangement of equal layer thicknesses. As a check, $m = 1$ returns a centre-line strain that depends only on the average of Λ_T and Λ_B. Assuming that the latter can be different values (each E is the same but each α is not) there is also some curving, but in essence ϵ and κ are decoupled in this special case.

Replacing each ϵ term in Eqs. (21.2) by Eq. (21.5), the axial force in each layer is now expressed in terms of κ alone. Substituting them into the left-hand side of M, Eq. (21.4), it can be shown after some heavy algebra that:

$$ M = \frac{E_B wt^3}{96(1+m)} \left[\kappa \left(1 + 14m + m^2\right) - \frac{24m}{t} \cdot (\Lambda_T - \Lambda_B) \right]. \tag{21.6} $$

This is not the most obvious expression, again, so we test for special conditions. Note, however, the dimensional consistency between all terms and that we have a relationship between M, κ and Λ, which becomes our generalised Hooke's Law for bending of a bimetallic strip.

When M is zero, heating is unencumbered by applied moments, and a positive curvature emerges from the right-hand side provided $\Lambda_T > \Lambda_B$: a larger expansion of the top layer causes the strip to curve downwards, which we expect. Despite there being no overall bending moment, the stress resultants can be shown to be non-zero: as we surmised initially, heating alone is not a stress-free process.

When $m = 1$, Eq. (21.6) is simplified greatly:

$$ M = \frac{Ewt^3}{12} \left[\kappa - \frac{3}{2t} \cdot (\Lambda_T - \Lambda_B) \right]. \tag{21.7} $$

If there is no difference in thermal strains, we arrive at the familiar Hooke's Law for simple bending of a rectangular cross-section, with $wt^3/12$ the usual second moment of area. Otherwise, a positive moment-free curvature always follows, equal to $(3/2t) \cdot (\Lambda_T - \Lambda_B)$.

21.2   'Controlling' Displacements

We have consciously tackled this problem using stress resultants acting on discrete layers rather than by integrating continuous variations of stress, and their turning effects, through the thickness for force and moment equilibrium. As noted, there is no need to transform the cross-section, and we accord structural insight which simplifies the formulation even though the final algebra for M is wieldy – but unavoidable, no matter our approach.

The coupling in the above expressions underlines the *controllability* of performance: that we can tailor the shape (via κ) whatever the loading (from M) by setting the difference in Λ. This principle precisely underpins the bimetallic strip being used as a thermostatic switch and, more recently, so-called *smart materials* such as piezo-electric and shape-memory alloys, in novel shape-changing structural applications.

To demonstrate as such, the strip is behaving as a tip-loaded cantilever in Fig. 21.2(d). A force F is applied upwards, and we seek the level of heating needed

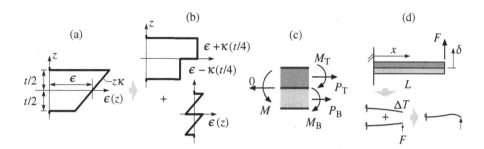

Figure 21.2 (a) General linear strain variation through the thickness of a bimetallic strip. (b) Separation of the distribution in (a) into purely extensional components, top, and purely bending, bottom, in each layer. (c) Moment equilibrium across an element of strip, showing the net forces and bending moments on the right side. (d) Tip-loaded and heated cantilever strip. Each effect on the displacements superpose to give the final displaced shape.

to *induce* zero tip displacement. We will assume that $m = 1$ for simplicity but that $\Lambda_T \neq \Lambda_B$: the cross-sectional geometry remains and E is the Young's Modulus.

The sequence of 'loading' does not matter: indeed, if heated first, the strip curves downwards and F can be applied to undo, or *block*, any tip displacement. The blocking force is, therefore, used as one measure of shape-changing, or *actuating*, capability.

We can superpose each loading event because we are dealing with small deflections and linearity throughout, in order to arrive at a result most quickly. For heating alone, the strip curvature is uniform where, from the Author's note, the tip deflection downwards is $\kappa(L^2/2)$ with κ here equal to $(3/2t) \cdot (\Lambda_T - \Lambda_B)$.

The upwards tip deflection is a standard case from Chapter 10 equal to $FL^3/3EI$. These two components must balance, which sets either:

$$\Lambda_T - \Lambda_B = (\alpha_T - \alpha_B)\Delta T = \frac{4}{9}\frac{FLt}{EI} \quad \text{or} \quad F = \frac{9EI}{4L}\frac{(\Lambda_T - \Lambda_B)}{t} \tag{21.8}$$

depending on which explicit variable you prefer.

For a sense of the scale of F or the level of heating, we can write in a different dimensional way:

$$\frac{F}{wt} = \frac{3Et}{L16} \cdot (\Lambda_T - \Lambda_B) \tag{21.9}$$

after replacing I. Both sides have units of stress and t/L is typically around 1/50. Thus $F/wt \approx E(\Lambda_T - \Lambda_b)/200$. Engineering materials typically yield at a strain of around 0.1%, which sets a limit of 0.2 for the difference in thermal strains for F approaching yield conditions – approximately so.

21.3   Final Remarks

The bimetallic strip is the simplest example of a controllable shape-changing structure. Fundamentally, the mechanical features – stress and strain, moments and curvature *etc.* – are coupled to a non-mechanical stimulus, in this case, a source of heating.

Other materials can be stimulated differently; for example, piezoelectric crystal materials shear in response to an electric field through its thickness (applied via surface electrodes).

But, unlike heated metals, piezo-materials operate in reverse: they offer an electrical *output* in response to mechanical loading. Such *electro-mechanical* coupling promotes increased functionality, for example, in applications dealing with energy generation.

Applying a pressure produces an electric field, but electric charge – the energy that can be stored – is only developed when the field is discharged to a resistive circuit and recharged again: we must load and unload the material in order to extract a supply of charge (for example, for harvesting electrical energy from a mechanical action).

Further Reading

There is a wealth of formal instruction on Structural Mechanics, Design and Engineering, which I have had the pleasure of reading, in full or in part. The reader is referred to the following for extra reading.

Beginners may be interested in reading:

J E Gordon, *Structures — or Why Things Don't Fall Down*, DaCapo Press, 2003

J E Gordon, *the New Science of Strong Materials — or Why You Don't Fall Through the Floor*, Penguin, 1991

M Salvadori, *Why Buildings Stand Up*, W W Norton & Company Ltd, 1991

J Heyman, *Structural Analysis: A Historical Approach*, Cambridge University Press, 1998

On aspects of general theory and analysis, see:

J Heyman, *The Science of Structural Engineering*, Imperial College Press, 1999

J Case, C T F Ross, H Chilver, *Strength of Materials and Structures*, Arnold, 1999

A C Palmer, *Structural Mechanics*, Oxford University Press, 1976

R C Coates, M G Coutie, F K Kong, *Structural Analysis*, Van Nostrand Reinhold, 1988

For details of plastic analysis, see:

J Heyman, *Plastic Design of Frames Volume 2: Applications*, Cambridge University Press, 2011

C R Calladine, *Plasticity for Engineers*, Ellis Horwood, 2000

Various design approaches are considered in:

W Addis, *Structural Engineering: The Nature of Theory and Design*, Ellis Horwood, 1990

M F Ashby, *Materials Selection in Mechanical Design*, Butterworth Heinemann, 2010

And for related topics involving structures, such as Aerospace- and Mechanical Engineering, see:

T H G Megson, *Aircraft Structures for Engineering Students*, Elsevier, 2012

R R Archer, S H Crandall, N C Dahl, T J Lardner, M Srinivasin Sivakumar, *An Introduction to Mechanics of Solids*, Tata McGraw-Hill, 2012

J L Meriam, L G Kraige, *Engineering Mechanics Volume 1: Statics*, Wiley, 2008

Index